HERMES, ECOPSYCHOLOGY, AND COMPLEXITY THEORY

Also by Dennis Merritt

Jung and Ecopsychology
The Dairy Farmer's Guide to the Universe Volume I
ISBN 978-1-926715-42-1

The Cry of Merlin: Jung, the Prototypical Ecopsychologist
The Dairy Farmer's Guide to the Universe Volume II
ISBN 978-1-926715-43-8

Land, Weather, Seasons, Insects: An Archetypal View
The Dairy Farmer's Guide to the Universe Volume IV
ISBN 978-1-926715-45-2

HERMES, ECOPSYCHOLOGY, AND COMPLEXITY THEORY

THE DAIRY FARMER'S GUIDE TO THE UNIVERSE VOLUME III

DENNIS L. MERRITT, PH.D.

fisher king press

Published simultaneously in Canada, the United Kingdom, and the United States of America by Fisher King Press. For information on obtaining permission for use of material from this work, submit a written request to:
permissions@fisherkingpress.com

Fisher King Press
PO Box 222321
Carmel, CA 93922
www.fisherkingpress.com
info@fisherkingpress.com
+1-831-238-7799

Every effort has been made to trace all copyright holders; however, if any have been overlooked, the author will be pleased to make the necessary arrangements at the first opportunity. Many thanks to all who have directly and indirectly granted permission to quote their work, including:

From *The Homeric Hymns*, 2nd ed., translated by Charles Boer, copyright 1970 by Spring Publications. Used by permission of Spring Publications.

From *Hermes the Thief: the Evolution of a Myth* by Norman O. Brown, copyright 1969 by Vintage Books. Used by permission of Steiner Books.

From *Dance Therapy and Depth Psychology: The Moving Imagination* by Joan Chodorow, copyright 1991 by Routledge. Used by permission of Taylor and Francis Group LLC-Books.

From *Archetypal Psychology: A Brief Account* by James Hillman, copyright 1983 by Spring Publications. Used by permission of Spring Publications.

From *Hermes Guide of Souls* by Karl Kerenyi, translated by Murray Stein, copyright 1976 by Spring Publications. Used by permission of Spring Publications.

From *Hermes and his Children* by Raphael Lopez-Pedraza, copyright 1977 by Spring Publications. Used by permission of Spring Publications.

From *The Sacred Prostitute: Eternal Aspect of the Feminine* by Nancy Qualls-Corbett, copyright 1988 by Inner City Books. Used by permission of Inner City Books.

From *The Black Goddess and the Unseen Real* by Peter Redgrove, copyright 1987 by Grove Press. Used by permission of David Higham Associates.

L. H. Stewart's figure of the Archetypal Aspects of the Self appearing on p. 142 of *Archetypal Processes in Psychotherapy* edited by Nathan Schwartz-Salant and Murray Stein, copyright 1987 by Chiron Publications. Used by permission of Chiron Publications.

From *C. G. Jung: His Myth in Our Time* by Marie-Louise von Franz, copyright 1975 by Hodder and Stoughton. Used by permission of Inner City Books.

CONTENTS

The four volumes of *The Dairy Farmer's Guide to the Universe* offer a comprehensive presentation of Jungian ecopsychology. Volume 1, *Jung and Ecopsychology*, examines the evolution of the Western dysfunctional relationship with the environment, explores the theoretical framework and concepts of Jungian ecopsychology, and describes how it could be applied to psychotherapy, our educational system, and our relationship with indigenous peoples. Volume 2, *The Cry of Merlin: Jung, the Prototypical Ecopsychologist*, reveals how an individual's biography can be treated in an ecopsychological manner and articulates how Jung's life experiences make him the prototypical ecopsychologist. Volume 3, *Hermes, Ecopsychology, and Complexity Theory*, provides an archetypal, mythological and symbolic foundation for Jungian ecopsychology. Volume 4, *Land, Weather, Seasons, Insects: An Archetypal View* describes how a deep, soulful connection can be made with these elements through a Jungian ecopsychological approach. This involves the use of science, myths, symbols, dreams, Native American spirituality, imaginal psychology and the *I Ching*. Together, these volumes provide what I hope will be a useful handbook for psychologists and environmentalists seeking to imagine and enact a healthier relationship with their psyches and the world of which they are a part.

My thanks to Craig Werner for his comprehensive and sensitive editorial work, and to Tom Lane, Rinda West and Rosalind Woodward for their constructive comments.

x

To the Great Goddess in her many forms

Hermes

"For all to whom life is an adventure—whether an adventure of love or of spirit—he is the common guide."

—Karl Kerenyi

CHAPTER 1

Hermes and the Gods

Hermes may be *the* Western figure for establishing a mythological base for ecopsychology. He illuminates the processes and perspectives that will allow us to develop a psychology of ecology and an ecology of psychology; a psychology of depth, imagination, myth and symbolism in relation to each other and to the environment. Hermes is the god of psychologists and businessmen—two important elements of ecopsychology—and offers a link to Native American spirituality and its connection to the land. He is also important in male sexuality, male spirituality and issues of the body-mind connection. Most significant is Hermes' role in establishing communications and relationships across *all* levels; between the gods and goddesses, between the divine and human, between the living and the spirit world, and between humans and animals. As god of synchronicity, Hermes is about relationships between particulars and levels not encompassed by Western science— electronics, the mind, the imaginal world, organic and inorganic, etc. He fulfills his roles to a large extent due to his personification of the revolutionary mathematics of complexity theory that describes the creation, dynamics and evolution of complex systems from the inorganic realm through human dream activity. Hermes portrays in symbolic and mythic form the human experience of the mathematics of complexity theory.

Hermes/Mercury was the god of alchemy that became Jung's main symbolic system and the historical context for his "confrontation with the unconscious." (volume 2 of *The Dairy Farmer's Guide*) The archetypal energies represented by Hermes were deeply experienced by Jung beginning with his childhood nightmare of a giant phallus on a golden throne. Jung felt we should become familiar with our forgotten or neglected Western roots before wandering off to other parts of the globe for enlightenment. Greece, the source of Western culture, is our mother/father lode here. Three main factors contributed to the dynamism of this exceptionally rich culture that flourished around 500 BCE.

1. The Greeks in the culturally fertile Mediterranean basin (including Egypt, Crete, and the Middle East) created their own unique synthesis from psychic and cultural "trade." This included a melding of nomadic sky-god cults that infiltrated the earthbound, agricultural Great Goddess religions. (Graves 1960, p. 17ff)

2. The culture developed on the cusp between nature religions that experienced a plethora of spirits in the natural environment and the laying of the Western foundations for writing, science, mathematics, logic, literature, history and art.

3. Greek society had evolved from tribal culture to kingship to the first forms of Western democracy. (Brown 1969)

Greek culture left us with an abundance of writings, images, marbled archaeological sites and a trail of other cultural influences. The rediscovery and reemphasis of our Greek roots in 16th century Italy helped birth the Renaissance which brought Europe out of the Dark Ages. Returning to our pre-Christian roots can help us decipher what went wrong with our philosophical, scientific, educational and religious systems that have led to the current dysfunctional state of our relationship to the environment.

The Greek pantheon presents a Western version of fundamental energies and images that are still alive if we have an eye for them. Stories and images of gods and goddesses help bring to consciousness the basic, archetypal dimension of the psyche and the physical world. Gods and goddesses express the basic metaphors, perspectives, and ways of being in the world. Many Greeks knew the gods did not literally exist but were more than real because their powers and influences could be felt as those things that "come over" a person, just happen to one, motivate or depress, etc. The best and deepest expression of the *reality* of human existence is therefore through myth and poetry about the sacred. (Hillman 1975, p. 13-17)

Each divinity is a sort of essence, "a kind of spiritual condensation," of a realm of being. (Kerenyi 1976, p. 47) Each had a dimension of the real world as apprehended from their perspective, "forming a unified totality in its own right." (p. 3) One could think of each species of plant and animal in this manner. Conceiving the human psyche in this polyvalent, polymorphous way is inherently ecological and lays the foundation for an ecopsychology.

Each god and goddess is like a comprehensive idea or worldview, a mythic gestalt that encloses us and creates our world; an archetype in other words. (Kerenyi 1976, p. 46, 47, 55) Consider how the Cartesian idea of the world as a giant mechanical clock affects our way of "seeing," responding to, and being in the world. That world is made clearer, more conscious, when we call the name of the god or goddess active at the moment and label the experience in their honor. One must know mythology in order to do this. The transpersonal dimension is brought into life when one becomes conscious of what timeless human drama one is enacting in one's personal way. (see Appendix H: Archetypal Psychology and Aphrodite as the Soul of the World)

Hermes is an unusual god in that he personifies the unconscious propensity to produce gods and goddesses. He goes to a level beneath the gods in that he is the source of the gods, in that sense more like the Tao. "As the basis of understanding the world, he is also idea, though one we have not yet fully grasped," as the famous mythologist Karl Kerenyi described it. (Kerenyi 1976, p. 55)(n 1)

Hermes was one of twelve gods in the Greek pantheon and the last to join that hallowed group. His roots go to the phallic core and foundations of the Western psyche. The most basic form he was worshiped as was the Herm, an upright stone that served as a phallic monument. Nature itself offered this object and "the stone pointed to a direct experience of something divine." (Kerenyi 1976, p. 78) Later, stone pillars called "Herms" marked property boundaries and stood in front of Greek houses for protection. (Bolen 1989, p. 163)

The quadratic form of the phallic shaped stone or wooden Herm originated in the Greek province of Arcadia where Hermes was especially honored. (Kerenyi 1976, p. 68) Kellene was the name of the Arcadian mountain where his cult was located. The Hermes cult may be a very ancient one, there when the Arcadians arrived, and the Greeks adopted this ancient indigenous god and gave him a Greek name. (Lopez-Pedraza 1977, p. 1) The Athenian Hermes had a head atop a quadratic pillar bearing an erect phallus, the pillar portion being called the herm. The head symbolized Hermes' self-knowing, self-conscious nature (Kerenyi 1976, p. 78) while the quadratic ground plan is an archetypal expression of totality, incorporating the chthonic dimension of life. (p. 68)

Hermes was also associated with stone heaps, the *hermax*, that were placed along roadsides as landmarks for travelers. (Kerenyi 1976, p. 78) Passers-by would add stones, asking Hermes as Lord of the Roads for protection on their journeys. (Bolen 1989, p. 163; Lopez-Pedraza 1977, p. 3) Graves were often placed beside roads and the stone mounds frequently marked the graves.

CHAPTER 2

The Genesis of Hermes in the Individual and in our Culture

The *Homeric Hymn to Hermes* reveals how the Greeks wrestled with the link between human and animal, chthonic and divine in the sixth century BCE. (n 2) These core issues are alive for us today in an in-depth approach to ecopsychology. The Hymn opens with Hermes' conception:

> [Maia] awesome, lying with Zeus
> She kept away from
> the wonderful company of the gods,
> and lived in a shady cave.
> Here the son of Cronus had
> the nymph with beautiful hair,
> in the early hours of evening,
> while sweet sleep held
> the pale arms of Hera,
> and where no man
> and god could see. (Boer 1970, p. 18)

Zeus, ruler of an Olympian pantheon of gods and goddesses, whose activities and intrigues are the mythic dimensions of human experience, sows the seed of Hermes' existence on one of his many sexual escapades. Zeus was not drawn to the known and bright heights of Olympian goddesses, but to a nymph bound up with the Arcadian landscape where she was a type of primordial mother-daughter Goddess. (Kerenyi 1976, p. 19) Hermes will become the only *nymph's* son to have a permanent place in Olympus and "the only one who really knows what underlies the word 'nymph.'" (Lopez-Pedraza 1977, p. 90) Kerenyi sensed in Hermes the essence of the pre-Olympian world, the world of the Titans. (n 3) Hermes entrance into the Olympian world of Zeus integrates Titan energy into the Olympians. (Kerenyi 1976, p. 17, 18)

The integration occurred through the conception and development of Hermes. It reconnected Greek culture with its chthonic roots, to the aspects left out or left behind from its Olympian heights. Maia shunned the sacred congregation of the gods and chose to live in a cave. Perhaps she did not feel at home with the Olympians, did not feel accepted and respected? She may have been unable to relate to those worldviews so unlike or threatening to her.

It was Maia's beautiful hair that attracted and aroused the ruling god's fantasy. Hair, growing out of the head, is associated with thoughts and fantasies. This is a fantasy Zeus can allow himself to entertain under the conditions that no god or mortal will see him: it is night, his jealous wife is asleep, and he is in a dark cave with a nymph who doesn't associate with the gods. Hermes is conceived in the sacred vessel so deep within ourselves it is beyond all the structured ways of looking at and being in the world—beyond even the gods and religions. In that space is complete freedom to play, imagine and fantasize with the uninhibited innocence of a child. It's the space where the active, generative, primal phallic energy (Hexagram 1, The Creative, the yang source in the *I Ching*) conjugates with the unconscious matrix that evokes and attracts it in a secretive and seemingly forbidden (Hermetic) way (Hexagram 44, Coming to Meet). (n 4). Sexual fantasies are included in the Homeric story line. This is the fluid source and uninhibited, uncensored domain from which dreams arise. It is the domain of beginnings and origins; the depth and realm that sages, shamans and seers are cognizant of, giving them oracular powers (Hermes is god of oracles). They see things at their source long before they have developed enough to manifest in the light of consciousness. Dreams serve the same function, and it is Hermes who brings sleep and dreams.

The gestalt of Hermes' conception provides root associations of Hermes with night and the "carrier of dreams," or a "secret agent" usually portrayed as a youth. Hermes is Lord of Roads, protector of travelers, and god of psychologists because he is associated with journeys into and out of the dark depths of psychic existence. To evoke Hermes' name is to ask help to see life as a journey; to loosen psyche's bindings in concretisms, literalisms, rigid positions, collective opinions and dogmas. Hermes' realm is the road and pathway as an existential journey: it is not simply in getting from point A to point B, but one's journey through life. The world-of-the-road has Hermes for its god and is the mythic base of America's love affair with the automobile. Beyond all boundaries, Hermes realm is the source of life's journey. Hermes is constantly in motion and his ability to volatize (dissolve forms) gives him access to everything (Kerenyi 1976, p. 14, 15):

> With companions of the journey [with Hermes], one experiences openness to the extent of purest nakedness, as though he who is on the journey had left behind every stitch of clothing or covering...Journeying is the best condition for loving. The gorges over which the "volatized one" passes like a ghost can be the abysses of unbelievable love affairs...[where] no chance exists for standing on firm ground. (p. 14)

A psychotherapeutic journey may take months before people can reach this place of Hermes' conception within themselves, months before trust and a relationship is built with the therapist that can contain the energies beneath defenses and character armor. Transformative therapy commences when the analysand can begin to play, to "see" metaphorically, to create.

The secret affair in the cave is more than just a sexual romp for Zeus: it fulfills a wish for Zeus and through its fulfillment his "mind" or "insight" achieves its end. (Kerenyi 1976, p. 20). This associates Hermes' phallic nature with light and consciousness. (Jung 1961, p. 13) What is generated in the dark as "just" a thought, idea or fantasy can bring a whole new consciousness into being: (n 5)

> when the mind of great Zeus
> was near to completion...
> and he was bringing to light again
> all his great works,
> she produced her child (Boer 1970, p. 18)

After a powerful experience in one's life that journeys the soul along, we have to carry and protect our fantasies in a private, moist, nourishing darkness. A painter cannot receive projections and critiques of her unfinished works. Therapy must be private and confidential. An analyst has the analysand look back nine months in their lives after a dream of giving birth to a child: usually a soul-moving event happened at that time. Psyche keeps its own time. Privately we have to come to terms with new conceptions before the world sees them; ideas must evolve and develop before exposure to Darwin's cruel light.

Hermes was born in the morning on the fourth day of the month, one of his several associations with the number four. (Kerenyi 1976, p.

22) Four was also sacred to Aphrodite with whom he is closely linked. Jung emphasized the association of four with totality—the spirit incarnate.

Constant motion and activity like the unconscious is associated with Hermes, a trait revealed at his moment of birth:

> For after he jumped down from
> the immortal loins of his mother
> he couldn't lie still very long (Boer 1970, p. 19)

CHAPTER 3

The Power of Music

Hermes leapt up and went in search of Apollo's cattle, only to find a mountain tortoise at the entrance to the cave. Hermes exclaimed:

> What a great sign,
> what a help this is for me!
> I won't ignore it.
> Hello there,
> little creature,
> dancing up and down,
> companion at festivals,
> how exciting it is
> to see you. (Boer 1970, p. 20)

Hermes admires its "kaleidoscopic" shell as he carried "this lovable toy" back home, "stabbed out the life of the turtle," and in the twinkling of an eye came up with the idea of making the first Greek lyre. (Boer 1970, p. 20)

This episode reveals varied aspects of Hermes' character, beginning with his "meeting and finding" nature—the happiness and riches associated with accidental findings. *Hermaion* is the Greek word for windfall. It is also the name of the roadside offerings left at the herms that were stolen by hungry travelers in the spirit of Hermes. Kerenyi sees the Hermetic sanctity of "the accidental find...seized as a thief" in every business undertaking that is not unscrupulous. Here finding and artfully clever thievery occurs in "a no-man's land, a Hermetic intermediate realm that exists between the rigid boundaries of 'mine and yours.'" (Kerenyi 1976, p. 24) It is the clever execution of a deal within the context of the saying, "a fair exchange is no robbery."

In dealing with the turtle we see the quick, inventive mind and restless nature associated with Hermetic energy, a mind that delights in the joy of discovery. A Hermetic moment is in a lucky find, an intuitive insight, a synchronistic event, a transformative happening. It can recognize the beauty and potential in an object or situation. A Hermes' move can take those awkward, rigid, defended turtles of our lives and transform them into the gaiety of a festival and an energetic dance— probably after a few drinks and to the right musical accompaniment. Kerenyi sees in Hermes "the...freedom of soaring flight for which he gives wings to even the most sluggish souls." (Boer 1970, p. 58)

Hermes under the name Mercury became the god of alchemy, and the transformation of the tortoise into the lyre epitomizes the alchemical process. The slow, heavy tortoise symbolizes the lead in our lives, the raw material, the *"prima materia"* one starts with. In Chinese alchemy, the tortoise was regarded as "'the starting point of development,' the beginning of the spiritualization of matter." (Chevalier and Gheerbrant 1994, p. 1018) Hermes told the tortoise: "Alive...you're good medicine against the pains of black magic. But dead, dead you'll make great music!" (Boer 1970, p. 20) This is the essence of the alchemical work: antidotes carry an element of the poisonous characteristics in them. (Chevalier and Gheerbrant 1994, p. 1018) The feminine symbolism of the tortoise can be negative—its chthonic nature, involutive chaos, regression and sexual attributes, but properly worked with, the introversion and regression to the source within the protective container of the shell can lead to spiritual transformation. (n 6)

There are two other aspects of Hermetic inventiveness. On one hand Hermes can personify the creative source of scientific musings in areas of basic research (on the decline in America). The second aspect is that "necessity is the mother of invention." Hermes' quickness in immediately recognizing that the turtle can be a help to him is "just as a thought runs quick through the heart of a man whose troubles pile up and shake him." (Boer 1970, p. 21) This aspect of Hermes' nature can be high in basic survival values–one of many traits he shares with the Greek god Eros. (n 7)

Hermes reveals the ambivalent nature of a divine and a cruel Titanic self in his dealing with the turtle. He can "see through" the living turtle to a divine instrument that can be constructed from its dead form. This "seeing through" is the divine standpoint offered the spectators of a Greek tragedy that allows them to participate in the play with a penetrating vision of what's really happening. (Kerenyi 1976, p. 26) It is the perspective of an analyst or therapist who can "see through" to the dynamics of a neurotic complex and sense the mythic story line the patient is living.

Kerenyi highlights Hermes' roughish and Titanically cruel nature that laughs at the turtle while clearly pointing to the violent death he will inflict upon it to transform it into a lyre. The divine side shines through as well, "for out of his victim's death he conjures music, the unique way for a mortal human to transform the harshness of existence into Phaiakian mildness." (Kerenyi 1976, p. 26)

As Hermes started to play the lyre he created from the turtle,

> It sounded terrible!
> The god tried to improvise,
> singing along beautifully,
> as teen-age boys do,
> mockingly, at festivals
> making their smart cracks.
>> He sang about
>> Zeus, the son of Cronus,
>> and Maia in her beautiful shoes,
>> how they talked during their love affair,
>> a boast about
>> his own glorious origin.
>> And he honored the servants
>> of the nymph
>> and her magnificent house. (Boer 1970, p. 21, 22)

Boastful, roguish, adolescent males embody an aspect of a Hermes' mentality that knows no shame. A mentality that can sing of the pillow talk of his parents on the night of his conception; unbounded sexual fantasy indeed! A mythical world away from Pauline Christianity and a virgin birth via a spirit! That prominent erection, the head atop the ithyphallic herm, symbolizes a self-consciousness that can go back to the primal sexual scene: No Freudian trauma here. Some Buddhists believe a soul is drawn to reincarnate by seeing the conjugal act of its parents-to-be. Kerenyi said *The Hymn to Hermes* "may be called a highly literary monument to phallic shamelessness." (Kerenyi 1976, p. 29) His chthonic, phallic nature is associated more with Titanic, pre-Olympian roots (p. 32) while "finding and thieving in the realm of love are also Hermetic traits." (p. 28)

Another dimension of Hermes is revealed in his shameless song. The Greek word *genee* is rendered as "begetting" in Hermes' "tale of his begetting." It involves the abstract origins presented by mythology, "the basic reasons for everything that exists" or will exist. "In the genealogy, ancestors of 'famous names'" (*geneen onomakluton*) occupy the place of origin, as emergence proceeds forth from the primordial depths. The genealogy turns the great original mythic theme into a family tree." We see this in Genesis in the Bible and in ancient Chinese culture with the reverence of ancestors and the mythological founders of dynasties. "The family tree must begin, of course, with the earliest gods...[Hermes'] impudence is the consciousness of his own origin and reason for being, an unbroken and linear consciousness of his development which is...a...characteristic feature of Hermes." (Kerenyi 1976, p. 30) In psychoanalysis, this is seeing through to the archetypal base of the complexes, naming the eternal gods or goddesses as the root/ originating elements at play in the patient's life. (see Appendix H) In ecopsychology and deep ecology, it is discerning the archetypal base of our dysfunctional relationship with the environment.

With his singing and playing Hermes honored the feminine, the mother source; the rich unconscious matrix symbolized by the "magnificent house in a cave." He also honored the servants of his mother and becomes the god of servants because he serves the gods. (n 8) He is not the heroic super hero—he never engaged in battle. (Kerenyi 1976, p. 8) Hermes uses craft, wit and music to avert and subvert the heroic stance, as we shall see in his dealings with Apollo. Jesus assumed the servant role when he washed the feet of his disciples, and rejected Satan's offer to rule the world.

CHAPTER 4

Hermes and the Cows

Evening approached as Hermes finished playing the lyre, and he had a hunger of pre-Olympian dimensions. The Greek phrase *kreion eratidzon* translates as "being extremely greedy for meat"; the phrase applied to a lion's hunger in the *Iliad* and denoting a craving aspect of Hermes' nature. (Kerenyi 1976, p. 32) (n 9) Hermes bounded out of the cave a second time, "working on a shrewd trick in his head, like those done by robber types who operate at this hour of the dark night" [Hermes as thief and highwayman]. (Boer 1970, p. 22) His plan? To steal 50 of Apollo's cows: Apollo, an Olympic god, Zeus's favorite son and right hand man! Here's a baby with balls! That's Hermes.

Far-sighted Apollo, god of a distanced and reasoned approach and logical thought, could look into the very mind of God to see the structure of the universe and the fates of men. Apollo the archer was straight as an arrow whose trajectory could be calculated with Newtonian linearity. Here we have the Greek god of science, the Apollo space program, and "scientific" university psychology programs. Is this not perfection? Yet the Greek genius perceived an archetypal need to complement the Apollonian worldview. It conceived of a Hermetic mind-set that ingenuously steals in the dark from Apollo's vulnerable cow side; that sees in the dark beyond the bright, dominant worldview. This is what Zeus had "in mind" when he "stole off" in the dark to mate with Maia.

Hermes made off with 50 of Apollo's cows leaving behind the black bull and four dogs. The bull and the dog have associations with archetypal masculine, so Hermes' choice emphasizes his countering Apollo through Apollo's vulnerable feminine/cow side. Fifty was an important lunar (archetypal feminine) number for the Greeks, there being 50 lunations between Olympic games and 50 priestesses in various Moon-Goddess cults (with associations to the erotic and the human). (De Vries 1974, p. 182)

The worldview from Hermes' perspective is like seeing life from the reverse, backward side—the unconscious, also associated with the unpredictable nature of Hermes' steps. Hermes displayed his tricky nature by turning the front hooves of the cattle backward and the back hooves frontward. He bounded along backwards on sandals made by twisting together tamarisk and myrtle branches tied to his feet. The sandals were precursors to his winged golden sandals by which he effortlessly covers great distances in his messenger role. These sandals were "indescribable, unimaginable, they were marvelous creations," demonstrating Hermes' inventive and enchanting nature. He "avoid[ed] a wearisome trip by wearing such original shoes." (Boer 1970, p. 24) One association with shoes is one's standpoint in life. Our standpoint/ viewpoint has a lot to do with whether or not we see life's journey as being wearisome or we bring something creative and enchanting to it.

The psychic ancestress of the Hermetic standpoint is his mother's "beautiful shoes" which would represent the roots of a tree for a tree nymph. Myrtle is a plant associated with death: we are dealing with "swift as death" Hermes here and that which destroys forms. A quickness and cleverness associated with death transforms one of Aphrodite's animals (the tortoise) into a musical instrumental and root-shoes into indescribable sandals. Stagnant, restricted old forms must be killed off for new forms to be created, establishing creativity's link with death. (n 10)

Hermes drove the cows to a secret stable where he invented the art of fire making by rubbing sticks together. Then he dragged out two cows and "filled with great power" threw them onto their backs and sacrificed them. (Kerenyi 1976, p. 26) (n 11) He divided the finest meat into twelve parts as gift offerings to the gods, counting himself as one of the gods. The delicious smells tempted Hermes to eat the sacred meat:

> His noble heart
> persuaded him, however,
> not to let them pass
> down his own divine gullet,
> though he wanted to,
> badly. (Boer 1970, p. 27) (n 12)

The fire myth that a culture lives out will produce two diametrically opposed results depending on the myth. Fire from the god Hermes is associated with arousal and sacrifice of the flesh as an equal offering to *all* the gods, thereby being accepted by the gods. Such an act avoids hubris by recognizing and honoring powers bigger than the ego—the transpersonal domain of the gods—which the ego falsely believes it can control. The West is enacting a Promethean version of the fire myth. Prometheus *stole* the fire from the gods to aid humankind. He himself was not a god and honor was not given to the gods. Prometheus was more akin to Titan stock operating more out of a power and force modality: human hubris was increased as a consequence. The gods' punishment for Promethean-aided humans was Pandora's box. Each god and goddess made their contribution to creating an irresistibly beautiful woman who would seduce and deceive the human race, loosing the world's evils when she opened her infamous box. (n 13) This is played out in sexy commercials used to sell powerful automobiles that pollute the planet and contribute to climate change. The automobile is an engineering and scientific triumph and feverish minds in top ad agencies develop million-dollar-thirty-second commercials for the Super Bowl to pedal these ingenious creations.

The combination of Titanic *and* Olympian elements in Hermes' theft and sacrifice must be emphasized. The bloody offering and inordinate hunger for meat firmly establishes Hermes' connection to the Titanic, chthonic realm. In the end, symbolically taking part in the offering and not giving way to his greed for meat links him to the more distant Olympian realm. (Kerenyi 1976, p. 32, 33) Hermes is one of *the* gods for the connection to our fleshy "animal" lusts *and* an imaginative dance with the flesh. Only by being so close and so tempted can the symbolizing dimension of the psyche be fully engaged and the psyche transformed/volatilized by the subsequent sacrifice of immediately living out desires and impulses or a willfulness to force one's control over a situation. One must get intimately close, then be "initiated" by death (myrtle branches), a night journey, sacrifice, or a backward walking way—it can feel like a crucifixion. The sacrifice allows one to "see through" to the particular god or goddess that is arousing the flesh and not blindly living it out from a personal perspective. The ultimate is dying before your physical death—Plato's definition of philosophy.

This god is close to my heart. Recognize him every time you get a Big Mac attack or smell that big juicy steak sizzling on the backyard grill. Even though divine, Hermes barely resisted gobbling down his tender piece of meat. The meat sacrifice was presaged by Hermes' accidental discovery of the turtle—a mythological primal animal with a connection to Aphrodite—his killing it and turning it into a work of art, and the mysterious transformative dimension of music. (Kerenyi 1976, p.

25) Blues music is a perfect example of turning life's lusts and suffering into an art form that adds a transcendent dimension as it deepens an experience and generates an aesthetic richness in life. (n 14)

Hermes has introduced a new consciousness, a new divine standpoint: "divine theft."(Kerenyi 1976, p. 34) Titanic theft uses power and might to achieve its ends. Hermes' divine Olympian nature allows him to hold back, a backward way; a distancing from the immediate instinctual aided by an objective detachment associated with death. This changes Titanic nature into ingenious charm and forgoes violence. (p. 33) Inventiveness and animated swiftness replaces violence. The sophisticated pathway of the symbolizing dimension of psyche is engaged. The pathway of divine, Hermetic theft is ingenious, swift as death, enchanting, easy; like Hermes backward steps on his tamarisk-myrtle sandals. (p. 34) "Material" gets "volatilized" just as meat is given up in a burnt offering: the concrete and literal finds its mythic, spiritual core. Theft becomes more associated with merchants and businessmen than with military conquest, petty larceny or sandlot bullies. There is a childhood innocence to Hermes' thieving, which is ultimately done to affect a transformation. (n 15)

Hermes modeled the honoring of all the gods and goddesses, recognizing the unique aspects of each. An ecopsychological perspective sees the unique traits and position of every individual or species, not allowing any one individual or group to dominate. Hermes is the first to proclaim his membership among the Olympians and the last to join the group. By becoming its twelfth member he turns the pantheon into a symbolic whole and links the Greeks back to their roots at many levels. (Kerenyi 1976, p. 33)

After his amazing theft Hermes quietly returned home at dawn (a transition point associated with him). His manner of re-entering the cave metaphorically describes a Hermes' entrance into our life, a Hermes' move in the psyche:

> [He] slid in through an opening
> > into the room,
> > like a breeze in autumn,
> > like a mist. (Boer 1970, p. 28)

This is Hermes; a hidden presence, a spirit, volatized and invisible.

Hermes innocently curled up under his baby blanket while holding his lyre in his hand. But his mother wasn't born yesterday; she knew he was out all night plundering:

> Get out! Your father made you
> just to be a headache
> to gods and men! (Boer 1970, p. 29)

Hermes answered:

> Why, I shall be engaged
> in the greatest art of all—
> always concerned for you,
> of course, and for myself.
> We're not going to stick around here,
> as you want, the only two
> among all the immortal gods
> without any gifts,
> without even prayers!
> ...
> As for honors,
> I'm going to get in on the same ones
> that are sacred to Apollo.
> And if my father wouldn't stand for it,
> I'll still try,
> I'm capable certainly,
> to be thief number one. (Boer 1970, p. 29, 30)

And if Apollo comes searching for Hermes, he threatens to go and steal from Apollo's house.

Hermes, the bastard son full of chutzpah and wit, is determined to get his share of the pie. This is the mentality of a second son, of a self-made man who rises from obscurity, not always by honest means, something like a Joe Kennedy. Hermes is the active force to bring about the recognition and redemption of what he and his mother represent.

Talk is cheap: we haven't heard from Apollo yet. The next morning Apollo realized his cows were missing and knew that the thief was a son of Zeus by divinating from the overflight of a long winged bird (a crane, Hermes' sacred bird). (Graves 1960, p. 67) Apollo was amazed with the reversed cow tracks and astounded by the foot print monstrosities of the quick-footed herder that went from one side of the road to the other.

He was in a bad mood; "He covered his broad shoulders in a dark cloud" (Boer 1970, p. 33), and rushed off to Maia's misty cave. Baby Hermes tried to hide himself deep under the baby blankets and pretended to be asleep. Apollo knew Maia and Hermes and intrusively searched the back rooms of the house, using a silver key to open storerooms. He then threatened to attack Hermes in his cradle and

> throw [him] into black Tartarus,
> into hopeless darkness.
> What a terrible end! (p. 36)

Hermes would forever remain underground without hope of rescue by his parents, leading the dead around "who drift about insubstantially with a 'frail buzzing' or 'chirping'...[in] the 'likenesses of living creatures.'" (Kerenyi 1976, p. 37)

Hermes proclaims complete innocence, that as a day-old baby he doesn't even know what cattle are, he'd only heard of them. He just wants the tender care of his mother and adds:

> And it would be a big surprise
> for the gods: a baby,
> just new-born
> who could walk right in the door
> with a herd of cows. (Boer 1970, p. 37)

He offers to swear an oath of innocence on his father's head. And while he is pleading his innocence he is acting like someone listening to a lie! That makes Apollo laugh! Even the god is impressed with this incredibly clever trickster, calling him, "The Prince of Thieves"; a name by which he will be honored among the gods. Hermes brings a sacred dimension to thievery.

Apollo then seizes the baby and hoists him up. But Hermes intentionally and indecently released a stomach rumble and suddenly sneezed. This so disturbed Apollo he threw Hermes to the ground and caustically demanded that Hermes lead the way to his cattle. Hermes continued to proclaim his innocence to Apollo the "most violent of all the gods." In anger they walk off with each other to settle this in front of Zeus. Hermes realized "[he] had come up against someone very smart." (Boer 1970, p. 40)

Archetypal cleverness and thieving clashes with archetypal intelligence and rationality—and it's a standoff. We see the dark side of

Apollo surfacing as "the most violent of the gods"—and this is in a pantheon that includes the god of war! We've all seen the type—the intelligent, rational, ordered man who has been crossed: watch out. Last century we saw this at a national level—the most cultured, scientifically advanced Western nation, Germany, produced the Nazi beast. This was Jung wrestling with the Dionysian and dark side of Eros within his Germanic ancestry; the repressed Wotan that was the archetype of the Nazi beast and possessor of the German psyche. (Jung 1961, p. 234, 235, 318; CW10, sect. III) *This* is a problem that Greek genius is mythologically addressing in Hermes' story.

Why was Apollo so upset by Hermes' sneezing and his intentionally rumbling stomach? The myth is extracting Hermes' realm from Apollo's, symbolized by stealing Apollo's cows—archetypically feminine energy. Sneezing and rumbling of the stomach, a "nervous stomach," are controlled by the autonomic nervous system. This "vegetative" nervous system controls those automatic, deeply seated responses such as heartbeat, digestive movements, unconscious breathing rate, etc.—*basic* stuff. When we feel deeply anxious or are unconscious about something bothering us, our stomach may rumble. A more severe form of this is colitis. It is one thing to sneeze and display deep anxiety with a rumbling stomach, which the Greeks could read as omens; it's another dimension to *intentionally* cause it. We're talking about being tuned into the body at the level of automatic, inherited processes. These are states of consciousness achieved by yogi masters who can control their heartbeat and blood distribution in their bodies. This is symbolized by the statue of Hermes with a knowing head atop a phallus. Link this back to another Eastern association—Hermes' consciousness of his amorous parent's pillow talk at his conception.

Hermes' "evil belly-tenant" places him at the opposite end of the pole from "the mortally clean God Apollo." Hermes' realm is far, far from Apollonian consciousness, indeed, it feels very threatening to it. The Greeks created a place for it in the Western psyche. The range of the Hermetic world extends from the indecent to the light, delicate and dignified, pleasing the gods of the upper and lower regions. Hermes acceptance into the Olympian circle showed that "even the lowest is not unholy." (Kerenyi 1976, p. 39)

Apollo is so upset because he is losing an unknown something. His father has created a monster whose second act in life is to steal from him. Everyone knows Apollo is Zeus' favorite and closest to him. What's going on here? Let's see how Zeus handles these disturbed sons.

CHAPTER 5

On Trial Before Zeus

Chuckles break out among the assemblage of incorruptible gods as they anticipate the trial of the century between Golden Boy sun god and a baby cattle thief. God of Light, God of Truth, Silver Bow Apollo, the Archer (strikes from a distance) presents his case first:

> Father, you're going to hear now
> a rather difficult story—
> you who charge me alone
> with being the greediest
> person for loot. (Boer 1970, p. 41)

Golden boy is quite greedy, narcissistically inclined, maybe heartless in his thieving/looting. He proclaims Hermes to be an incomparable cheat and trickster.

Apollo's description of Hermes' movements and how difficult it is to track him is the perfect metaphor for what the old alchemists (and modern psychologists) wrestled with in dealing the hidden depths of the psyche. It is difficult to find even a trail to begin with. Apollo says," I had to search hard in many lands to find him." Referring to the "double tracks" of the cows, Apollo exclaims how fantastic and disturbing it was, "the work of a powerful demon!" The Hermetic moves in driving the cows were undetectable, "impossible" from Apollo's mind-set to imagine: "It was marvelous! It was as if he had used little trees for feet." (Boer 1970, p. 42)

Sandals, like "little trees for feet"—*The Odyssey* says the magic powers of plants are familiar to Hermes. (Kerenyi 1976, p. 34) Hermes' connection with the vegetative realm imaginally provides the chthonic roots of Greek culture and ecopsychology. Hermes' mother was a tree spirit noted for her "beautiful shoes" (Boer 1970, p. 21) and her "beautiful hair." (p. 18) The crown of a tree is the most beautiful, noticeable aspect of a tree and Maia's hair is what attracted Zeus. The crown performs the miracle that feeds all of life, receiving and converting pure solar energy into the chemical bonds in the sugars manufactured by the leaves. The carbon for the sugars comes from the leaves' absorption of life's metabolic waste, CO_2. What is not noticed and appreciated in a tree are its roots which perform the absorbing and supporting function. Roots live in the dark, moist domain of accumulated death and decay (the topsoil). They plunge deep into the purely inorganic mineral domain, absorbing water to provide the H_2O part of the photosynthetic equation.

Photosynthesis

Solar energy

$$6CO_2 + 6H_2O \longrightarrow C_6H_{12}O_6 \text{ (glucose sugar)} + 6O_2$$

Within the context of a single tree, water and the supportive and absorptive functions of the roots would be associated with the feminine; air (in this case the CO_2 part of air) and light are absorbed by the leaves in the crown and are more associated with spirit and the masculine. Hermes is like the trunk of the tree that connects the upper and lower domains. At the cellular level he is like the phloem and xylem cells that carry fluids up and down the tree.

The squirrel is the messenger communicating between the upper and lower realms of The Teutonic World Tree; the center of the world and a Teutonic Self image. The squirrel is related to Hermes ability to navigate between realms. Trees are Self images throughout world cultures because they extend across three domains (heaven, human and earth), are durable (associated with wisdom and eternity), have "character" (think of a gnarled old oak) and are of impressive size and stature.

Hermes made sandals out of tree branches and later sprouted the binding willow thongs on his feet into a maze (another root image). I mentioned earlier the association of the myrtle wood of the sandals with death; the volatilization/spiritualization of the earthy/chthonic. Through myth and music Hermes facilitates this process of distilling the essence and archetypal dimension from the mute vegetative and animal domain. Hermes' winged helmet would be related to the spiritualized essence of the tree crown.

Apollo brings up Hermes' association with night and the underworld, saying that upon returning home,

> he lay down in his cradle
> just like the black night itself,
> in the darkness of a misty cave.
> Not even an eagle,
> With its sharp eye,
> could have spotted him
> down there. (Boer 1970, p. 43)

Hermes' terrain is not Apollo's turf. The far-sighted, distanced approach to life is not Hermes—he's up close and personal, deeply rooted; the animal intelligence of dark, chthonic, body regions. His realm is not clear or logical and is often difficult and frustrating to follow. It is available if one is open to it, and by evoking Hermes' guidance on life's journey.

Hermes lays it on thick when it is his turn before Zeus. He says he doesn't even know how to lie! He tells Zeus how Apollo forcibly ordered him to talk and didn't even bring any gods as witnesses or observers (the Greek roots of our Bill of Rights?). Apollo even threatened to throw this poor, poor baby into Tartarus. Then Hermes says,

> Believe me—
> you who have the honor
> of boasting that you are
> my father—
> I didn't take his cattle home,
> though I do want to be rich.
> I didn't even step over
> our doorstep. (Boer 1970, p. 44, 45)

This is the Hermetic approach: be daring, be confident, be cocky ("Oceans Eleven" is an archetypal Hermes' movie). He reminds Zeus, in the middle of a civil war between two brothers, that Zeus has the honor of boasting that he is Hermes' father! He tells a Hermes' version of truth—a half-truth: he never *stepped* over his doorstep; he *bounded* over it. Hermes adds a great oath of "NOT GUILTY" and closes with a tug on Zeus's heart strings, reminding him how Zeus "help(s) us youngsters" against the stronger. (Boer 1970, p. 45)

And then, the *pièce de résistance*: Hermes winks! After putting on a performance that would put any courtroom lawyer to shame, he lets Zeus know it was all a lie: Hermes—god of rhetoric and persuasive speech. Hermes is not concerned with exactness and facts as an Apol-

Ionian type would present a convincing case. Hermes is about playing cleverly on people's emotions, doing whatever is necessary by whatever means to present a convincing argument and move the psyche, the jury, the consumer. This is the art the Greeks are honoring here. Americans saw this archetype on TV for the nine months of the O. J. Simpson trial as the best lawyers money could buy argued a case of passion and revenge.

Zeus' response to Hermes' song and dance—and wink? He burst out in a great laugh! Zeus had witnessed smoothness and trickery raised to an art, a sacred form, and that is to be revered. It is laughter that brought acceptance of hermetic energy into the Greek pantheon, demonstrating the importance of humor. But now, the real challenge is presented. The differences have been delineated and are combative. Zeus *orders* both of them to try to reconcile.

The story evolves as Zeus commands Hermes the Guide to lead the way to the hidden cattle, without any further mischief:

> Zeus merely nodded,
> and the noble Hermes obeyed him.
> For the mind of Zeus
> who carries the aegis
> persuades easily. (Boer 1970, p. 46)

The first step in reconciliation is to have a Self figure strong enough to set limits and initiate movement toward union. Hermes' energy, like phallic energy, can be impulsive, possessive, tricky, seductive and self-centered: think teenage male here. Hermes' Roman name is Mercury, the name of that fascinating, shimmering metal that is liquid at room temperature and can shatter into a thousand droplets and meld back together. It symbolizes many of Hermes' traits. Changeability and unpredictability—elements of creativity and inventiveness—can dissipate into a mirage of possibilities if nothing is followed through, if definite directions are not established. A father figure, a mentor, is often necessary for a Hermes type to succeed. Laws and systems are necessary to limit the excesses of *laissez faire* capitalism and big business tycoons. The opposite, Super Ego extreme is the state dominated economy of the old Soviet Union that completely stifled creativity, individualism, boldness and risk-taking. An educational system with too much structured learning (and testing) kills off Hermes; so does a psychology dominated by reductionism, causalism and statistical methods.

CHAPTER 6

Delineating Apollo's realm from Hermes' realm

So off go "these two charming children of Zeus" and Hermes releases the cattle from the cave. Apollo gets another shock: he sees the cow skins that Hermes stretched out to dry after butchering the two cows for a sacrifice. Hermes' power, even in infant form, scares Apollo who fears what it will be like full grown. He anxiously tries to bind Hermes' arms and feet with powerful thongs of willow:

> Those he put on his feet, however,
> suddenly started growing
> down into the ground,
> twisting together,
> and easily tangled up
> all the wild cattle there—
> thanks to the schemes
> of tricky Hermes.
> Apollo was shocked!
> Then the powerful Argiephontes
> looked up and down,
> suggestively, a fire
> twinkling in his eyes...
> ...he wanted to hide. (Boer 1970, p. 47)

In Hermes' pedigree is Atlas, father of his mother and the Titan who holds up the world. Intellectual, Apollonian, objective types can feel threatened by physicality and the raw, brute strength in the animal nature of the body. They fear this aspect of the hermetic domain and may seek to limit and constrict it. Apollo's unconscious cow-nature has become even more split off as it went from domestic to wild cattle that end up being tied up and fearful.

Causing the willow thongs on his feet to grow reinforces Hermes' connection to the vegetative realm of life and his influence over it. This ties us back to the tree sandals and their symbolism. The hermetic action associated with cattle is now seen by Apollo, unlike the thieving that occurred in the dark. Apollo is now directly and fully conscious of the power of the hermetic realm and fearful he may not be able to restrict this domain so foreign to him. He sees that any attempts to tie this energy down will result in further entanglement and restriction. This is the position of the Medical Establishment and the Religious Right at this moment in time. The "alternative health methods" and "New Age" spirituality are proving too fruitful to ignore and too popular to constrain. This is not to say there aren't a lot of charlatans and flakes in the "new" movements: Hermes, Guide of Souls, is also known to lead people astray. Too much experimenting with mercury by the alchemists looking for spiritual transformation led to brain poisoning and the Mad Hatter phenomenon (hatters used mercury to tan leather; insanity from mercury poisoning was an occupational hazard).

Hermes' activity with the willow thongs reminds us he is the god of binding and unbinding, the god of keys and locks. Enzymes are the Hermes of the biochemical realm. All life is made possible by the catalytic activity of enzymes that bring two forms together to create a synthesis or to break up an old form. The Apollonian approach in searching for what binds one and locks one up can become angry and intrusive as shown by Apollo's manner of searching Maia's back rooms of the cave and opening boxes with his silver key. It is like a therapist intrusively pushing too hard to overcome a patient's resistance in an attempt to get to the bottom of things and meet an HMO timetable. But that which binds you can also release you. The most basic archetypal illustration of this aspect of Hermes is that he is the god associated with cutting the umbilical cord after childbirth; the symbolic and literal separation from the Mother.

How does Hermes handle Apollo's attack? He doesn't get angry, reprimand, curse or make fun of him. He furtively looks about, eyes twinkling, and wants to disappear: Hermes is a master at hiding and disappearing. Just when you think you've got him...a furtive glance, a suggestive movement—and he's gone.

Hermes' next act begins the Greek lysis of our stuck and fearful Western position. It was foreshadowed by Hermes' first act, even before stealing from Apollo. It's his greatest gift; he steals only to bring soft light to what is being overlooked or neglected. Hermes performs the important ecopsychological function of transforming our individual and cultural neurosis only after we've hermetically moved down into the deep affective realms and the level of the autonomic nervous system. Has the society of humans been shocked enough by the atrocities we commit in the environment like the massive BP oil spill in the Gulf of Mexico? Once Hermes has got us consciously at that deep level, he transforms by reconnecting us to our archetypal roots and myths in a manner literally and metaphorically expressed by the mysterious power of music. This move of Hermes is the lysis of the dilemma and "*It was very easy for him to soothe the Archer.*" He began to melodically play his lyre and

> it sounded marvelous!
> Phoebus Apollo
> was delighted, and
> burst into laughing.
> The lovely sound
> of this divine voice
> went right to his heart,
> and a sweet desire
> transfixed his spirit
> as he listened.
> > The son of Maia,
> > playing his lyre so charmingly,
> > took courage, and stood
> > on the left
> > of Phoebus Apollo.
> > Suddenly he started playing the lyre
> > louder, reciting a prelude—
> > and the sound accompanying him
> > was lovely—
> > about the immortal gods
> > and the dark earth,
> > how they were at the beginning,
> > and what prerogatives each one had.
> > And the first of the gods
> > that he commemorated with his song
> > was Mnemosyne, Mother of Muses,
> > for the son of Maia
> > was a follower of hers.
> > And all of them,
> > all the immortal gods,
> > according to age

and how each one was born,
the glorious son of Zeus
recited, singing them all
in order, playing his lyre
on his arm. (Boer 1970, p. 47, 48)

Humans inherited the ability, personified by Hermes, to produce divine sound and mythic expression that can transform even an uptight and fearful Apollonian type. The Greek muse is asking us to imagine that first moment in human history when the sound of an instrument as beautiful as a lyre was marvelously played. That is a Hermes moment. This is deep, fundamental, vegetative-level stirring of consciousness symbolized by the ithyphallic stone Herm topped by a knowing head on the Athenian statues.

Chinese sages in the classic text, *The I Ching*, commented on the mysterious power of music over 2500 years ago, roughly about the same time as the origin of The *Homeric Hymn to Hermes*. In hexagram 16, Enthusiasm, it is written:

> Music has power to ease tension within the heart and to loosen the grip of obscure emotions. The enthusiasm of the heart expresses itself involuntarily in a burst of song, in dance and rhythmic movement of the body. From immemorial times the inspiring effect of the invisible sound that moves all hearts, and draws them together, has mystified mankind.

> Music was looked upon as something serious and holy, designed to purify the feelings of men. It fell to music to glorify the virtues of heroes and thus to construct a bridge to the world of the unseen. (Wilhelm 1967, p. 68, 69)

Classical music, opera and classic popular songs like '60s Dylan continue to carry our mythic stories and archetypal themes.

Hermes also personifies the continuous myth-creating activity of the human psyche. The hermetic instinct is to "know" the archetypal roots and mythogenic level of the human psyche and discern what archetypes/gods are active in the moment. It knows all the gods, their attributes, their genesis and stories and presents in mythic fashion the eternal "story" of our connection to nature. A hermetic move has courage in its approach because of this knowledge and possesses a sense of timing that knows when to move in and up the ante to influence the disaffected.

The first of the gods Hermes commemorated was Mnemosyne, Mother of Muses, whom Hermes follows. (n 16) The Great Goddess Mnemosyne was a Titaness and one of the wives of Zeus. This Greek personification of memory was described as ever-flowing and gushing forth through the poets. (Kerenyi 1976, p. 31) She is like a daimon of Hermes' fate; a fate to be possessed by memory carried as "the inherited knowledge of all primordial sources of being"; a type of consciousness of a more spiritual and deep psychological nature. (p. 32)

In oral cultures, memory included all knowledge, practical know-how, history and mythology. Ginette Paris, in her beautiful book *Pagan Grace*, describes Mnemosyne's attributes:

> Mnemosyne uses the structures of narrative, epic, song or myth to preserve remembrance. She loves repetition, rhyme, rhythms and the strong images that hold narrative together...Her goal is to evoke rather than to describe. In this style of memory the factual and the symbolic, the historical and the mythical, "real" events and "imaginary" happenings are all tangled up inextricably...It's an active memory which breaks into consciousness through archetypes, dreams and myths, fantasies, symbols and artistic work...Sometimes it pays close attention to what was intensely lived, while at other times it selects events which seem unimportant but are endowed with a depth that can neither be denied nor explained.
>
> [Mnemosyne is memory] not just of the past, a taped recording; it is constructive, evocative, poignant, and the beginning of musing as Mnemosyne was the mother of Muses. (Paris 1990, p. 121)
>
> The individual Greek who knew Homer by heart didn't have to "search his memory" to remember a certain passage; the words came effortlessly to his lips at the right moment as if a voice had whispered to him, thanks to Mnemosyne... Passages "learned by heart" should be available just as my native language is available, immediately and directly. (p. 131)

In our culture it is the filmmakers, advertising agencies and spin doctors who operate in Mnemosyne's realm. Myths and archetypes are their domain and "they count on our shared background to move us." (Paris 1990, p. 126)

Therapists honor Mnemosyne. Gestalt psychologists "emphasize that memory must necessarily adjust, distort and transform recollections to serve human personality in an environment that is both complex and changing." (Paris 1990, p. 132, 133) Body therapists can release emotions and distinct memories when massaging tight backs and tense muscles as if memory were stored in the body. Clients tell their subjective stories and subjective histories for healing effect; the "talking cure" or "memory cure." (Paris 1990, p. 124)

Myths are like therapeutic stories that affect whole groups, stories that carry collective meaning:

> Not just narration and remembering what happened, it's also a fundamental tale outside of time which tells us about something that happened once, is happening now and will repeat itself, always different and always the same. Hesiod says of Mnemosyne that she has knowledge of "what is, what will be, what was." Memory the myth-maker weaves the fabric of our lives. (Paris 1990, p. 124)

Mnemosyne counters the Freudian reductive method by helping us find meaning and a transpersonal bearing that sustains and contains us by seeing the part we are playing in a particular mythic drama that might otherwise be called pathological. A woman might be deeply depressed because she is incarnating Persephone in Hades at this period of her life. A psychiatrist objectively assessing her situation with a battery of psychological tests might label her clinically depressed, "not functioning normally," and give her anti-depressants, thus aborting the soul transforming possibilities inherent in the myth. Being able to relate to the mythic dimension of human experience and see ourselves as part of a bigger picture is an important aspect of ecopsychology. (see Appendix K)

The Hermes myth is about the significant cusp between oral and written tradition/literacy. The book metaphor for memory is to willfully search our mental data banks to find the right information. This more objective, distanced approach is ego-directed and Apollonian. The memory associated with Mnemosyne is like a divine female voice that whispers to us out of the experiential fabric of our collective lives. (Paris 1990, p. 129, 130)

Computer memory can encompass all book memory and begin to mimic the brain's ability to make original associations between bits of information. This includes computers being able to judge the importance of the information and weigh it accordingly. (Paris 1990, p.136) But computers lack Mnemosyne's nuanced capacity to lie:

> If I ask my mother what I was like as a child, she can give a clever answer, lie a little, or half-consciously disguise the truth, having considered my present-day concerns and guessing why I ask the question. She can play down or amplify character traits which became dominant later on. Here Hermes is the ally of Mnemosyne. If a computer is programmed to lie, it can only say wrong things or randomly reverse certain affirmations, which is very different from a falsehood inspired by Hermes, the patron of liars. (p. 137, 138)

The Greeks could not distinguish between the goddess of Truth and the goddess of Deception. They saw them not as inevitable antagonists, but as

> two indispensable elements of knowledge, for all truth contains some mystery. The role of Deception is important to Truth because it supports ambiguity; it prevents any one belief from becoming so absolute that knowledge itself is threatened...It would weaken knowledge to eliminate the ambiguity that Deception provides. (Paris 1990, p. 138)

> As the nine Muses were her daughters, so the arts present this ambiguity of truth and illusion...Deception is necessary to give birth to Mnemosyne's daughters, the arts. (p. 139) (n 17)

The relationship of Truth and Deception was expressed in Chinese culture by the yin-yang image. No culture and environment will be healthy until the yin and yang elements are in dynamic, healthy inter-relationship—a process facilitated by art, music and fiction. Within Apollo's framework, his inner relationship to Hermes is indicated by the black dot in the white yang. Hermes is our way in and down, backing us into his realm with Hermetic moves, facilitated by Mnemosyne.

Western males may find it easier to relate to the myth of Hermes than Western females for whom a goddess myth may appear to be more "natural." Hermes is actually an androgynous character closely related to very yang Apollo *and* the archetypal yin energy of cows and the great primal mother goddess Mnemosyne. Hermes is androgynous because he represents the *boundary* between opposites; the transition space and the dynamics within that space.

Apollo triumphed over Hermes' version of history beginning with the ancient Greek founders of the discipline of history. The principle of Herodotus and Thucydides was that history begins where myth ends.

"The Muses, autonomous divinities in the oral culture, become second class vis-à-vis Apollo, archetype of literacy," Paris notes. (Paris 1990, p. 133, 134) Historians try to be objective by focusing on facts, significant dates, etc., cutting through superstition and dogma. These Apollonian minds denigrate metaphorical thinking, looking for the truth and not myth, facts and not fiction. (p. 133) Apollo may have become the archetype of literacy, but the Muses, the daughters of Mnemosyne, unconsciously influence the worldviews of the historians. The "truth" in history depends on who is writing the history books, choosing the "facts" that fit their theoretical and cultural perspective. The French write a very different history than the British about the battle of Trafalgar, and the American Indian version of our cowboys-and-Indians history is finally being heard. The "myth" about what happened in the West is still a "reality" in the American society, culture and politics. Ronald Reagan rode the cowboy myth all the way to the presidency, and George W. Bush didn't miss many photo ops on his Texas ranch.

The Muse is also active "in the scientific theory which organizes facts and observations in a way that reflects the scientist's own favorite fiction." (Paris 1990, p. 134) The natural world is so varied and complex that examples can be selected to support just about any theory. (Merritt 1988, p. 22-25; *A Jungian Bouquet* in preparation)

To return to our tale, we see that Hermes has just charmed Apollo with his new lyre and song, recounting the origin myths and stories of the gods and goddesses, especially honoring Mnemosyne. In primal tribes the shamans are the mythic repositories and tribal genealogists, personal and mythological.

Apollo can hardly contain his greedy nature as he covets Hermes' lyre and says:

> you who work so hard
> > on inventions,
> > companion of festivals,
> > this song of yours
> > is worth fifty head.
> > From now on
> > I think our differences
> > can be settled peaceably. (Boer 1970, p. 49)

He heaps praise on Hermes' singing, "this marvelous and fresh voice, which nobody else ever knew how to do." (Boer 1970, p. 49) He proclaims that this Muse for incurable sorrows gives one jolliness, love and sweet sleep—the revelation of Hermes' essence. (Kerenyi 1976, p. 26) Because of these skills Apollo offers to make Hermes "the famous and rich guide among the Gods," which will bring to him and his mother wonderful presents and glory among the gods. (Boer 1970, p. 50) Hermes becomes "the guide among the gods" because of his association with the mysterious and creative power of music to form and transform, his deep knowledge of the gods because he births the gods from the maternal matrix of the collective unconscious, and his artistic capacities associated with the Mother of the muses.

Hermes tells Apollo he can immediately have what he wants, and extends friendship to him; Hermes is a generous, friendly god. As carrier of the collective mythic memory of his culture, he proceeds to tell Apollo he knows exactly who he is, both by way of praising Apollo and to let him know he knows the score. He demonstrates why he's the god of diplomats, who share many of the same skills with good advertising personnel—they sense the basic cultural attitudes, their strengths and vulnerabilities, and manipulate them with finesse:

> But you know everything
> very well in your mind.
> You sit in first place
> among the gods,
> son of Zeus.
> You're good and strong.
> Wise Zeus loves you,
> and it's only right that he does,
> and he's brought you
> wonderful presents.
>
> They say
> you learned the honors due to gods
> and the oracles from Zeus,
> and all his laws.
> I have learned myself that you have
> an abundance of all these things
> And here too I know
> why it is you are rich.
> It's up to you
> to learn
> whatever it is you want.

And since the spirit moves you
to play the lyre,
sing, play it,
enjoy the fun
that you receive from me.
But give the glory to me,
friend. (Boer 1970, p. 51)

The Greeks through Hermes' speech are delineating an Apollonian realm of knowledge of the universe and the arts as that which is obtainable by will, applied learning and practice, formalities, intelligence, foresight and objectivity. Jung described the unconscious as being a spectrum from the somatic unconscious at one end associated with the body and trailing off into physiology and the psychic unconscious at the other end associated with imagery, concepts and an organizing center—the Self. (Schwartz-Salant 1982, p. 111-121) Apollo's knowledge is associated with the psychic unconscious and the Self as Zeus while Hermes is more associated with knowledge springing from the somatic unconscious and its relationship with the feminine matrix—the unconscious. Apollo had no clue about the source of Hermes' inventiveness and creativity—it was a mystery to him. Apollo wonders if Hermes was "born with a talent for this fantastic thing" or given a great gift and taught by someone? "Do you have to practice?" Apollo asks. (Boer 1970, p. 49) In Hermes' realm, things spring from the well of the collective unconscious rooted in nature itself—the creative force called spirit that is innate in the cosmos. The appropriate conscious response is acknowledgment, reverence and appreciation of this source. Identification with the source and thinking that consciousness is all there is leads to the hubris we see today. Scientists overlook the fact that their great thoughts and hypotheses "come" to them, "inspire" them; they don't think them out or think them into existence.

Great inspirations explode like a bombshell in the psyche. A raw and primal nature accompanies a holistic nascent form. It takes months and sometimes years of thinking and articulation, perhaps research and/or practice as well, to manifest the inspiration into an elaborated form. Hermes is that initial shock of revelation, given like a gift, a grace. Sometimes it's a false alarm starting one off on the wrong trail: that's Hermes too. It can be lost, like to a "thief in the night." The Greek genius is telling us to acknowledge and honor this force that can be symbolized by a phallic dance with the feminine depths of the psyche.

This is illustrated by a dream I had after an incredible flash of inspiration. I was reviewing an acquaintance's hexagrams at Bahnhof Enge in Zurich while waiting for the train. Suddenly I was overwhelmed by the thought of a relationship between synchronicity, the hexagrams of the *I Ching*, and evolutionary theory. My mind was a frenzied beehive of thoughts for a whole week, leaving me disoriented within and without. I developed these thoughts over the next nine months that culminated in my thesis at the Jung Institute in Zurich titled, "Synchronicity Experiments with the *I Ching* and their Relevance to the Theory of Evolution." The first or second night after the Bahnhof Enge experience I had this dream:

> I am in my maternal grandmother's kitchen in the early morning with my friend "Emma." We have just spent the night together. She says I owe her $10,000. I didn't realize any money was involved.

"Emma" was an intelligent, attractive, spiritual woman I had known some years before. I enjoyed sharing my early enthusiasm about Jungian psychology with her. We didn't have a physical relationship because she was married and I was in a committed relationship. The setting in the dream was the "realm of the mothers"; the old, ancestral feminine ("grandmother's place")—the collective unconscious. There I had a one-night stand with money involved: Hermes chasing the nymphs in association with a "lucky find," the "source," "stolen love" and "cheap sex." (Lopez-Pedraza 1977, p. 90-100) But this was not cheap; something of incredible value had transpired in our intimate relationship at night. This was symbolic of a connection/conjunction with the creative depths of the unconscious as is often personified in feminine form to a man (the anima). Emma appeared as my personal image of the maiden/nymph form of the Great Goddess. She felt the value was coming from her: I owed *her* money. (n 18)

Our intercourse, an exchange, was the Hermetic act in the dream, as was Hermes next act with Apollo. Hermes extolled the wonders of his lyre and generously gave Apollo his "clear voiced companion," telling Apollo he knows how to express himself "beautifully and in harmony." There can be an all-out, elated, physical aspect to Hermes as revealed by him telling Apollo to bring the lyre

> to some rollicking festival,
> to some pleasant dance,

even to all-out revelry!
It's fun day or night! (Boer 1970, p. 52)

He'll take the cattle offered by Apollo for the lyre and tells Apollo

And the cows will mate with bulls
promiscuously, and bring forth
an abundance of males and females.
It's not right then
for you to be violently angry,
even if you are a bit greedy. (Boer 1970, p. 53)

Hermes' sexual, phallic, chthonic nature shows through with the promiscuous mating fantasy and he closes with a little dig at Apollo's greedy nature. This lets Apollo know he's got him psyched up yet Hermes is generous with him. The Greek psyche continues to delineate Apollo's from Hermes' realm when Apollo then gives Hermes his shining whip, the symbol of shepherd-hood, giving Hermes charge of the care of cattle. What Hermes had to steal from Apollo is now consciously given to his domain and honored by Apollo; "And the son of Maia received it laughing." (Boer 1970, p. 53) (n 19)

As the son of a dairy farmer, I can see I have found my god. No farming is more intense and intimate with animals than dairy farming. (See "An Archetypal View of the Midwest Environment" in volume 4 of *The Dairy Farmer's Guide*) The human who demonstrates an unparalleled connection to the spirit of the herd animal is Professor Temple Grandin at the University of Colorado. She is autistic; so disconnected from humans that she has to study them like foreign objects and memorize how to act in social situations so she can appear "normal." She knows how fearful herd animals are; how easily they spook, how quickly they are frightened by noises. (Grandin 2005; Grandin and Barron 2005) A Hermes type could be cognizant of the deep and crippling fears in a person's psyche, a necessary sensitivity for self-psychologists of the Kohutian ilk, or any good therapist.

The two brothers enjoy the harp as they return to Zeus on Mount Olympus. Zeus is delighted and wants them to go further—to love each other. Hermes, being the more spontaneous and open of the two, decides to do so immediately. The proof that Hermes loves Apollo to this day?

> The proof of this is
> that he gave the Archer
> his lovely lyre
> and that he knew how to play it
> as soon as he picked it up. (Boer 1970, p. 53)

What a wonderful statement from ancient Greek culture. Having one of their most cherished musical instruments being played well is identified with the loving connection between the dramatic, yet complementary, opposites of Hermes and Apollo. Apollo is innately connected with Hermes; "he knew how to play [the lyre] as soon as he picked it up." But he couldn't have invented the lyre—it came out of a development of his shadow cow-side personified by Hermes. To an Apollo-type this appears to come out of the dark from nowhere in wondrously and relatively full-blown form. This intimate association between complementary opposites is played out in the sciences by the relationship between the scientific geniuses and the scientific herd; between the Newtons, Einsteins and Hawkins and the lesser lights who work and rework theory, in their own creative and often times routine and tedious way. They evolve the theories and diligently run the repetitious experiments to the point of statistical significance or insignificance.

Maybe Hermes can so easily relinquish his lyre because he knows he's perfectly capable of creating again—and he does, making the shepherd's pipe that can be heard from afar.

Hermes plays the critical role of *connecting* the upper and lower realms of the psyche. He is our guide for meeting Jung's ecopsychological challenge of uniting our cultured side with "the two million-year-old man within." His lyre in Apollo's hands symbolizes his upward connection to insight, intellect and refinement. The shepherd's pipe symbolizes his downward associations. The music of shepherds appears in the context of animal smells and roughing it, of being in the elements; elemental forces through which are herded life's vulnerable and resilient and indecent energies. Hermes can be an undignified god who allows therapists to befriend the undignified aspects in themselves and in their patients. (n 20)

Apollonian types are more cautious and paranoid than Hermetic types. Whereas Hermes was ready to love Apollo immediately, Apollo was fearful:

Clever son of Maia,
I'm afraid
that some day you're going to steal
my lyre and my curved bow.
You have the honor,
from Zeus, of being in charge of
exchanges among men. (Boer 1970, p. 54)

Hermes is the god of commerce and business—the realm of exchanges. It's his gift that allows one to see a real business opportunity, a niche no one has recognized, and immediately pounce on it when the time is right. It's the clever maneuvering for position in a market that could lead to big gains—or bust. Hermes is in the dickering over prices in an Eastern bazaar, where a seller's life depends on being able to size up a buyer and push their limits, best done in a friendly and somewhat playful exchange.

Apollonian types like things to be clear, clean, spelled out—no ambiguity. They are more risk-averse and invest more of themselves in carefully thought out positions. Hermes, by contrast, is associated with chance happenings and risking it all leading to great reward or great loss, making him the god of dice and lotteries. Hermes types are the caterpillars that give up everything for the possibility of becoming a butterfly. (n 21)

Raphael Lopez-Pedraza, in his seminal book *Hermes and His Children*, identifies an archetypal base of the split between Freud and Jung as the fearful exchange between an Apollo and a Hermes type. (Lopez-Pedraza 1977, p. 15) Jung asked Freud for an association to one of his dream elements when they were analyzing each other's dreams. Jung's hermetic style was to be open, explore, and approach another as an equal. Nothing should be too undignified when exploring the psyche as these two mythic adventurers were doing. Freud responded to Jung's request with a "look of the utmost suspicion." Then he said, *"But I cannot risk my authority!"* This is Apollo, fearful of losing his bow and lyre, i.e., his identity and position, in a hermetic exchange. This was totally foreign to Jung's hermetic nature: "At that moment he lost [his authority] altogether...The end of our relationship was already foreshadowed." (Jung 1961, p. 158)

Freud's model of the psyche was based on 19th century scientific causality. Neurotic problems were reduced to past childhood experiences that Freud presumed were of a sexual nature. Freud feared the "creeping mud of the occult"—the deeper levels of psychological experience in a domain imagined by Freud to be seething with primitive Id forces. Jung befriended those forces, saw there was gold in the dung, in fact the wellspring of many golden gods and silvery goddesses each with their own worldviews.

Apollo tells Hermes it would "make my heart favorable and friendly" if Hermes swore an oath to never again steal from him. Hermes agrees as well to stay away from Apollo's house. This makes Apollo feel so secure that he promised "in bond and friendship" there would never be anyone he would love more. (Boer 1970, p. 54)

An alchemical dictum is "separate and unite." Become aware of a difference between two things (consciousness), then recognize and develop their unique and separate natures. Proceed to unite them in the *mysterium coniunctionis*, the alchemical "marriage" of opposites, by discovering their underlying relationship and establishing a conscious *viva la difference* between the opposites.

Secure in his archetypal position/worldview, Apollo delineates Hermes' realm even further. First Apollo gives Hermes a marvelous, golden, triple-leafed wand for fortune and wealth that will keep him safe when carrying out the decrees of Zeus. With this staff, Hermes becomes the messenger for the gods, the wing-footed Mercury in Roman mythology (the FTD florist insignia with a staff). Hermes, the great communicator, was Ronald Reagan's archetypal base. Hermes gets this post because of his cleverness, quickness, cunning and versatility, boldness, diplomatic skills and friendliness. He communicates between gods and humans and also between the gods—this is significant. Each god and goddess has their own completely valid way of looking at and being in the world. They interact with each other, with archetypal passions and archetypal hostilities living out Shakespearean-type dramas. Hermes is the god to call upon, to recognize, when real communication is needed between Olympian-size oppositions; preferably before the opposing sides become so violent, hostile and intractable that they lose all possibilities of mediation. A disastrous end-game is inevitable when Mercutio (Mercury/Hermes) was killed in "Romeo and Juliet." (n 22)

On the more mundane human level, we see people desperately in need of Hermes who have a lot to offer, but can't communicate it to others. They may have great passion, love to give, talents to offer, sex to share—but they're stuck. A man without Hermes will have no seductive ability and his sexual expression will probably feel to a woman like an attack. (Paris 1990, p. 107, 108) In the *Homeric Hymn*, Maia had wealth in her cave, but it was hidden away. It took her clever and active Hermetic son to bring out her value to the world and release its energy into the Western psyche.

Hermes' communicative abilities are desperately needed in university systems. The gods and goddesses remain locked up in their ivory departments in paranoid defenses of their turfs. It will be quite a task for the ecological metaphor to "green" the university departments.

Apollo then tells Hermes that he can't have the oracles he has always been asking about. Apollo has sworn that only he among the gods shall know the divine secrets of the profound will of Zeus:

> I shall harm some men,
> help others,
> bothering many
> of the race of unenviable mankind. (Boer 1970, p. 55, 56)

Those who can interpret the cry and flights of prophetic birds associated with Apollo will profit, but they are on the wrong track if they question the oracles against Apollo's will in order to know more than the gods. Apollo does pass onto Hermes a very interesting oracle—a bee oracle—three Fates who are virgin sisters. The bee oracle is associated with the feminine and the three sisters are probably related by number to the Triple Goddess. Bees had a maiden's quality for the ancients (Kerenyi 1976, p. 41) and indeed all the workers in a bee colony are sterile females. The three sisters like Maia, Hermes' mother, *"apart they dwell,"* and their activity was not heeded by Zeus. They taught divination independently of Apollo, and Zeus didn't stop them. They get inspiration and pronounce truths when feeding on golden honey, but lead astray if denied this "sweet food of the gods." In other words, if the feminine is not acknowledged and nourished in a sacred way (seen as sacred), we not only lose its inspiration and prophetic attributes, but we are also led astray. This is a metaphorical statement of an ecopsychological concept about the importance of establishing a sacred relationship with nature and the chthonic realm. Hermes' connection with the bee oracle leads to his association with oracles and synchronicity where by meaningful chance the outer mirrors the inner. (see the end of chapter 3, "Planet of the Insect," in volume 4 of *The Dairy Farmer's Guide*) As a footnote, Apollo began practicing divination as a child around his cattle, another allusion to the association of divination with the deeper and archetypal feminine levels of the unconscious.

As an entomologist I find this very interesting. Bees' organizational skills and "intelligence" strikes the human psyche as spirit (order) in nature. We must have some knowledge of insects to decipher the bee oracle. Insects are very primitive animals, symbolically associated with the autonomic nervous system—a Hermetic association (cf. the stomach and sneezing oracles). There is an important association between a curious form of bee behavior and oracles as prophesy, foretelling, and reference to things not seen. Honeybees lapse into a frenzy when they find a rich new source of nectar. The bees that discover the new source of food communicate this to the other bees by doing a dance in the beehive, wagging their abdomen furiously while being closely followed by excitable bees. The messenger bee dances in a figure 8 pattern. The angle to the sun (bees see polarized light) of the crossing pattern of the dance tells the other bees the direction to fly to get to the nectar source. (fig. 1) The duration of the waggle dance along this orientation angle communicates how far it is to the source while the intensity of the dance correlates with the quality and quantity of the food. (n 23) The smell of the nectar and pollen on the messenger bee and taste of the regurgitated food tells the other bees the type of flower (these insects are extremely taste-sensitive and smell-oriented).

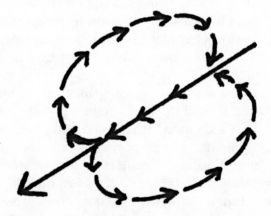

Figure 1. Arrow through the linear path of the bee
dance points in the direction of the nectar.

The other bees get excited, inspired over this gift, this find: this is their source of food for survival. Off they fly having been foretold how to orient themselves.

For artists and anyone engaged in creative activities, the bee oracle as inspiration are those exciting moments when it all comes together; new forms emerge that integrate discombobulated bits of existence. In that sense it is a revelation, a pro-spective, a foretelling of things to come. The whole body participates in the excitement and creation because thoughts and feelings are anchored in the body. But the period leading up to those moments is often uninspiring. Attempted work and products seem fruitless or misleading: the bees aren't feeding on the source of royal honey. Hermes then leads one astray if one attempts to force it.

In Dynamic Systems Theory terms, the conflict between the fragments leads to anxiety and depression, increasing the intensity until a phase transition is entered (the creative moment). This establishes a new and meaningful relationship of elements through self-organizing processes experienced as if one were receiving a revelation from an oracle. (see Appendices A and B)

Closer examination of bees and the bee oracle reveals why it is in keeping with the Hermetic realm. Apollo could prophesy from the flight of birds, something seen at a distance and associated with air; a more detached, far-sighted divination. When speaking of the bee oracle, Apollo addresses Hermes as a Daimon, which is close to the realm of mortals. Insects are very successful life forms, closely tuned to the physical universe of moisture, wind, light and temperature and so ancient that most flowering plants evolved together with insects. A symbiotic *relationship* exists between the attractive nectar producer and the pollinator. Flowers are like the primordial feminine that calls forth and attracts the primal inseminating male. Within the flower-bee dyad, bee activity is much like Hermes in his most chthonic, phallic form— flitting from flower to flower, spreading the inseminating pollen as he enjoys the flowery sweets.

Bees are full like souls full of enthusiasm when they discover their nectaries, the source material they condense into honey—the food of the gods. They delight in soothsaying (the bee dance) when enthusiastic; a soothsaying derived not from distancing, but intense, frenzied, sociable activity. The Greek word for their "swarming about" (*thaiosin*) means the swarming of the furious Maenads—female followers of the god Dionysus who went into a frenzy during their ceremonies. This relates to the inspiration and prophetic states attained with intercourse with the sacred prostitutes of the ancient Middle East. (see Appendix G: The Sacred Prostitute and the Erotic Feminine and Appendix H: The

Black Goddess) To the ancient Greeks, bees swarming about children lent them the gifts of the muses. Bees had soul and in antiquity were considered to be pure souls. (Kerenyi 1976, p. 41) Kerenyi points out that hermetic conditions of fullness in order to prophesy are expressed in pure intellectual form in Plato's Symposium. (p. 41, 42)

After bestowing the bee oracle, Apollo gives Hermes rule over all the animals—the instinctual realm of the chthonic if you will—no big surprise. This completes the demarcation of Hermes' realm in association with the chthonic domain and the vegetative unconscious. As mentioned before, Hermes is more associated with the somatic unconscious than the psychic unconscious; the more instinctual end of the instinct-spirit polarity. More importantly, he is the intermediary between the conscious and the unconscious, making him the god of depth psychologists. (n 24)

CHAPTER 7

The Sacred Phallus and the Guide of Souls

The final act in the Hymn is of particular significance. Apollo tells Hermes he is to be

> the only recognized messenger
> to Hades, who himself
>   never takes a gift from anybody.
>   This time, though, he will give him
>   a gift that is far from least. (Boer 1970, p. 57, 58)(n 25)

This is an allusion to Hermes as the guide of souls, the psychopomp; an important element of Hermes' nature that receives only this scant mention in the Hymn. To understand Hermes as psychopomp we must get to the root of the Hermes of life and death. This leads us back to the original, primordial, pre-Olympian Goddesses of the Near East. In those mythologems, Hermes arises not as the son of Zeus, but as "the first *evocation* of the purely masculine principle *through* the feminine." (Kerenyi 1976, p. 62, emphasis added) The seminal situation is the Great Goddess and the living primal Herm, in association with the primal waters, the area of becoming. In this context Hermes is the primal lover who is called forth or brought forth from the primal woman as her masculine counterpart or phallic servant-God. This is seen in ancient Greece with Hermes' many associations with bodies of water: associations with a spring with sacred fish in it; being honored in the vicinity of a swamp or spring; herm roadside markers that indicate to the wanderer the way to the next spring; or "according to one tradition, the Hermes statue—a very ancient Herm—...was fished out of the sea." (p. 64) (n 26)

Kerenyi sees this original Goddess as Brimo, the Great Goddess as the great evocatrice of northern Greece who contains within herself the germ-like form of "Demeter and Persephone on the one hand or with Artemis-Hecate on the other."(Kerenyi 1976, p. 63) "[Brimo] appeared in that elementary sort of maidenliness which does not fear the masculine as something lethally dangerous, but rather challenges, requests, and creates it." The masculine does not have an independent personality and functions as "a God-servant to the woman." "For the primal woman he was only an impersonal masculinity, almost a toy" and for primal Hermes the feminine was nothing more than an opportunity. (p. 63, 64)

Hermes is associated with the Silenoi; half-animal, phallic creatures, who chase nymphs and make love to them in grottos. While the Selenoi are "merely...the masculine counterpart and completion of the feminine nature-spirits, Hermes, in this relationship to the nymphs, is less one for whom the nymphs embody the Eternal-Feminine which he has to serve than one for whom they are an opportunity that he eternally masters." (Kerenyi 1976, p. 60) (n 27)

Kerenyi sees Hecate as the most Hermetic "of all the classical manifestations of the primordial Great Goddess who called Hermes into the world as the prototype of the secret lover." (Kerenyi 1976, p. 65). She sometimes took Hermes as her lover and sometimes the merman Triton, continuing the association of Hermes with bodies of water. (p. 64) (n 28) Like most ancient representations of the Hermetic essence Hecate was characterized by

> Associations with a kind of eroticism that one may find crass and vulgar and a connection to souls and spirits... Where Hecate ruled the world of northern Greece and Thrace in the form of "Aphrodite Zerynthia," it is precisely the crassest that is the holiest and most spiritual. (p. 65) (n 29)

Hermes priapic form and his epithet "having good fortune" in erotic affairs is shared with the lesser god Tychon in Attica. Tychon means "lucky marksman." (Kerenyi 1976, p. 68) "[Hermes] can be traced back to a masculine kind of life-force that remains very close to the feminine" but more active than the more constant feminine. It blesses the feminine "with itself, and with the continuance of its active nature, the child." The child is Eros in some myths or Hermes himself as in the Hymn. (p. 66)

Jung's phallic self-image is associated with a secretive, phallic nature mostly concealed and just emerging from the containing feminine. (see Appendix H in volume 2)

In the most ancient mythologems associated with stone Herms, the masculine aspect of the life force appears as congealed in its kernel (Kerenyi 1976, p. 66) and the kernel was associated with soul. (p. 71) The world of Hermes represented by the phallus as fertility symbol comes into being from the mysterious abyss of the active seed. (p. 82) At its deepest, timeless level it is the eternal source of all things, the source that has no beginning. (p. 73) It is pure being as the Hindu phallic Shiva is primal source being. (see chapter 2 note 11 in volume 1) There is a Gnostic link between Hermes and Osiris as creator of all things, whose phallus was proclaimed to be the object of the secret and extensive cult of Isis. (n 30)

The phallus is the primary masculine source of immortality because it is the eternal source of further procreation and life. (Kerenyi 1976, p. 71) "The original discharger of souls...remains forever the guide of souls, the messenger and herald between the realm of souls and the world of the born." (p. 75) (The feminine association with birth, death and resurrection is discussed in Appendix H: The Black Goddess.) The penis displays its uncanny ability to spring to life (erection, phallus), die and return to life again (res-erection). Kerenyi develops the association of ithyphallic Hermes with immortality through two of Hermes' connections; (1) Hermes' connection with the ithyphallic Kabeirian gods of the Samothrace mysteries who initiated Hecate into the underworld (n 31) and (2) Hermes as the reputed father of Priapus, a figure with a gigantic phallus. Hermes, like Priapus, was able to restore virility, and both have a similar relationship to death. Priapus guards gardens and graves and "wherever he is placed is...the place of life and death." (p. 69, 70) (n 32) Kerenyi comments, "If we conceive of the soul as masculine, as the eternal seed that is the begetter and procreator, it is also always what is begotten, at once father and son." (p. 79) (n 33) This is Father Time at the end of the year and the New Year's baby on January 1: time is continuous, it just has different phases.

This aspect of Hermes' nature is the archetypal root of Jung's childhood nightmare that associated the highest and most sacred value (Jesus, the temple and throne) with phallus, death and underworld. As a child Jung wondered about the subterranean god "not to be named" (Jung 1961, p. 13), "that alien guest who came both from above and from below." (p. 15)

The aura of brightness above the head of the giant phallus in Jung's childhood nightmare (Jung 1961, p. 12, 13) is also associated with Hermes. "[Hermes] is the designated bearer of all divine children, since he is the bringer of souls and of sun-children." (Kerenyi 1976, p. 89) A Hermes ritual, dedicated to eliciting the god's presence, had the handsomest young male carry a ram on his shoulders around the city walls. (p. 85) (n 34) Hermes acts as father and bearer of the ram (sun), but is not identified with the sun as source of the light. (p. 87) This association is in accord with an African tribe's focus on the sacredness of the moment of sunrise, the act or process of rising, rather than the sacredness of the sun itself. (Jung 1961, p. 267) (n 35)

The rising of something light-bringing from below, as a phallus rising from below, is associated with Hermes as an influx and invasion from the underworld, or "underworldly life," connecting Hermes with "serving spirits" and the "service industry." (Kerenyi 1976, p. 85) (see also note 8) A Hermes festival in Crete elevated the "low ones," the slaves, and had them served by their masters. At a Hermes festival in Samos people were allowed to steal and commit highway robbery. (p. 84) The Roman winter solstice festival of the Saturnalia also inverted the master-slave relationship:

> [Its] meaning was the strengthening of the weakest...That which hovers between being and non-being, seemingly powerless, repressed in servitude, reduced to the life in the nocturnal darkness of the seed, finds its way upward. Hermes, the psychopomp, also called Harmateus, the "soul carrier," guides, brings it back. (p. 85)

The sun is associated with the worldview Hermes mediates to our consciousness. On the "upper end" he's sworn to mutual love and respect for his brother, Apollo; the bright, far-sighted sun god who knows the mind of Zeus. Hermes as the inventor of language facilitates this upper end connection. The word *herma* is the verbal root for *hermeneia*, "explanation" and *hermeneus*, "interpreter"; a linguistic mediator also in a diplomatic sense. "By nature he is the begetter and bringer of something light-like, a clarifier, God of ex-position and inter-pretation... which seeks and in his spirit—the spirit of the shameless ex-position of his parents' love affair—is led forward to the deepest mystery." (Kerenyi 1976, p. 88)

Hermes association with sun and interpretation at his "upper end" linked with his connection to the nocturnal realm and the dream world makes him the god of dream interpretation. As the son of a nymph, he has a natural association with the realm of the mothers; the nocturnal and the instinctual realms. Jung realized that it is difficult to communicate with the dream world because,

> the concentrated brightness of our ego-consciousness has the effect of "dimming" the dream world, just as one scarcely sees candlelight when the electric light is turned on. The condition of unconsciousness keeps ideas and images at a much lower level of tension; they lose clarity and distinctness, their connections with one another seem less consistent, only "vague analogies." They do not seem to fit our logic nor to conform to temporal scales. (von Franz 1975, p. 94)

Ego consciousness is focused and concentrated, tending "to concentrate exclusively on adaptation to the circumstances of the present." This obscures or neglects unconscious material deemed inappropriate to adaptation, or unconscious energy too low or not yet ready for consciousness. A one-sidedness can develop and a sense of ego autonomy that lacks a sense of wholeness. (von Franz 1975, p. 96) Dream work counters this tendency and animates and inspires consciousness. (p. 94-96)

Jung's approach to dream interpretation occupies a subtle middle position (Hermes' domain) between the poles of spirit and matter, object and subject, causal and reductive, and final and prospective. (von Franz 1975, p. 97) Hermes is present when a good dream interpretation is made, establishing a link between consciousness and the unconscious; between mind, the body and emotions. (n 36)

Dream images (Hermes' domain) are amplified by looking at motifs in religions, myth and fairytales (Hermes' domain) and translating them (Hermes' domain) into modern psychological language. Interpretation thus allows connections or associations of the dream images to the lived psychic experience and an interpretation will have a clarifying, illuminating or enlightening effect. Intellectual interpretation alone is insufficient if the feeling value of the archetypal content is not experienced. (Jung 1964, p. 99)

Hermes' connection with the dream world, the animal realm and the deep unconscious makes him uniquely important to ecopsychology. His great mystery is manifested in

> the appearance of a speaking figure, the very embodiment as it were in a human-divine form of clear, articulated, play-related and therefore enchanting, language—its appear-

> ance in that deep primordial darkness where one expects
> only animal muteness, wordless silence, or cries of pleasure
> and pain. Hermes the "Whisperer" (*psithyristes*) inspirits
> the warmest animal darkness. His epiphany supplements
> the Silenos aspect of the life-source, in which the animal-
> istic factor within the Greek pantheon shows its presence,
> and within it forms a fundamental harmony and totality.
> (Kerenyi 1976, p. 88)

> The supreme knowledge of the Greeks...[is] that the
> Hermetic-spiritual aspect exists in friendly union with the
> animal-divine aspect. (p. 89)

These attributes of Hermes can be described in complexity theory
concepts as emergent phenomena within a symbolic system (see
Appendices A-F).

As we saw in the *Homeric Hymn* the link to the sexual and animal-
istic realm is facilitated by Hermes' association with music. Another
rendition of this theme is captured on a beautifully painted Greek vase
from the fifth century BCE showing Silenos, head of the Selenoi, and
Hermes together in the context of music being played. (n 37) On the
vase painting,

> The delicate figure of a deer between [Silenos and Hermes]
> hints at the untamed world which has been rendered
> tractable by Dionysian magic...[The] surface...is etched
> with the lines of eternal rhythm, the spirals...The facial
> features...tell everything—-the bestial yet grave face of the
> one and the super-humanly intelligent head of the other,
> and despite this difference the interfusion of their essential
> forms. Silenos has the lyre and lyre-pick of Hermes, while
> Hermes, behind him almost like a *Doppelganger* yet clearly
> marked by his winged hat and shoes, holds the Dionysian
> vessel of Silenos in his hand. They have exchanged roles,
> and this was allowed because at bottom, where Hermes is
> merely a Kabeiros, they have one and the same function:
> the conjuring of luminous life out of the dark abyss that
> each in his own way is. (Kerenyi 1976, p. 91)

CHAPTER 8

Hermes' Wand as a Symbol for Ecopsychology

Hermes' main attributes are captured in the symbolism of his wand—the caduceus. The wand had magical power augmented by being made of gold, itself associated with magical power. (Brown 1969, p. 16, 17) With the power to charm men to sleep and to make dreams come true, it was a "symbol of a golden age of peace and plenty." (p. 17) As the "ghost drawing" rod "it was an indispensable instrument in commerce with the dead" (p. 17) and was associated with Asclepius and healing. It was used to purify crossroads and protect the entrances to homes or symbolize prosperity and good fortune. (p. 6) The wand was also emblematic of the bard, priest and herald. (p. 28 note 38) Norman O. Brown sees Hermes with the wand as Hermes the Magician. (p. 17)

The essence of the psyche and relationships can be understood through examination of this symbol that I propose to be a symbol for ecopsychology. Revisiting our subject matter with another focus, that of Hermes' wand, gives us a better understanding of Hermes.

The original image of the wand was two serpents intertwined about a staff (fig. 2a). It appeared in the late archaic period as the girdle on a great primordial Goddess, the Giant Gorgo in Korfu, another form of the original Great Goddess Artemis. (Kerenyi 1976, p. 77) The two serpents represent any and all pairs of opposites, Winnicott's self-Other dichotomy, or any distinction between two things. (n 38) The distinctions can be between the different gods, between gods and people, mind and matter, spirit and soul, male and female, man and animal, within and without, truth and deception, non-existence and existence, temporal and eternal, life and death, science and religion, the individual and the collective, etc. The distinctions can be between different levels, such as the personal and the collective unconscious, and it is Hermes who connects the levels (metaphorically illustrated earlier as Hermes connecting the roots to the crown of the tree). Different levels of relationships within systems and between systems are a central concept in ecology and psychology.

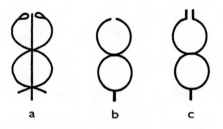

Figure 2.

The more abstract form of the caduceus is a figure 8 open at the top (fig. 2b). It both signifies and helps delineate differences. Hermes' domain is more precisely stated as the space between the two opposites as they face each other atop the figure 8. This space epitomizes the relationship between opposites, differences or levels that runs the gamut from antagonistic to loving. Emphasis on the space appears in certain depictions of the wand having parallel extensions from the top of the wand (fig. 2c)

Hermes' function as the god of the intermediate space, the space between opposites or differences, is rooted in the archetype of the trickster-magician in the autonomous clan and tribal unites of archaic Greek society. His main attributes are stealthy and secret activity, associated with mystery, with archaic Greeks not distinguishing between trickery and magic. (Brown 1969, p. 18-20) Strangers posed a threat to the autonomous family units requiring magic spells to safeguard interactions with them. Stone heaps associated with Hermes served as primitive boundary markers denoting the point of communication with strangers. "They were placed at the entrance of a house, where visitors were received; at crossroads or some other point on a road where strangers met habitually; in a forest or on some hilltop, both of which in a land like Greece constitute natural boundaries." Hermes was honored as "the guide" who presides over all comings and goings out, the "ambassador who protects men in their dealings with strangers." (p. 34) "Spell-binding, oath-taking, sexual seduction, and other forms of trickery reported in the myths of Hermes are in fact various manifestations of his magic power to control strangers," Brown noted. (p. 35) Hermes had the "skill of the oath" because of the magic binding power of the word. (p. 13, 14) The stealth power of the whispered word was associated with the magical seductions in love, linking Hermes with Aphrodite. (p. 14)

Hermes as the stealthy magician was a natural for being the god of boundaries and wastelands, also because he was the god of shepherds that grazed their flocks in the high wastelands far removed from the village (Hermes was not the god of cattle and cattle herding—cattle grazed at lower levels). "In the age of village communities the boundary was the scene of both pastoral and commercial activities; when commerce moved into the city, half of Hermes became exclusively rustic and pastoral, the other half became urban and commercial." (Brown 1969, p. 45) This contributes to Hermes being a good god for meeting Jung's challenge of uniting our cultured side with "the two million-year-old man within."

As god of the point or space between differences and what happens in such places, Hermes' stones marked property boundaries between what belonged to a person and what did not. He is associated with gates, which separate one side or property from another, and with door hinges, entrances and the socket or middle point about which things revolve. (Kerenyi 1976, p. 84) His times of the day are dawn and dusk— the transition points between night and day. Initiations and transition periods in life are his domain. At a deeper level, Hermes is associated with the boundary between existence and non-existence. As the guide of souls, the psychopomp, he brings souls into life, accompanies them on their journey, and transits them out of existence in their final journey out of life. Hermes can appear and disappear, like a person hovering near death, maybe regaining consciousness for one last time. Maybe not.

There were few temples to Hermes because, in an important sense, every household was his temple. Hermes was associated with "the most secret source and pivot point of human existence." This included the bridal chamber, bedroom and the center of the house as well as the hearth, where every now and then he appears in this "innermost nook." (Kerenyi 1976, p. 83, 84). Kerenyi writes, "Standing at the doorway, he indicates that here is a source of life and death, a place where souls break in, as though he were pointing out a spring of fresh water...Through Hermes, every house became an opening and a point of departure to the paths that come from far off and lead away into the distance." (p. 84) He guides souls within the house and without—the roadside Hermes aiding the soul's journey to and from home.

As the intermediary and the messenger Hermes is the god of ambassadors and diplomats. (see note 22) As master of the intermediary space, he has to be quick-witted, intelligent, cunning and versatile; have good timing, know each side's natures at a deep level, and have a sense of humor; be able to bend the "truth" to fit the circumstances, and can seize the moment. These traits also make him the god of businessmen.

Hermes in his messenger role needs all his diplomatic skills to establish communications and connections between the gods, and between the gods and humans. Every worldview is the domain of a god or goddess and trying to communicate between people with fundamentally different worldviews is difficult, cf. the conflict between the Christian and Muslim worlds. The clash of worldviews, or ignorance or repression of same, causes psychic dis-ease that expresses itself in physical and psychological symptoms in individuals and in cultures.

Hermes works best when one is open to him. This happens for most people only when they are *forced* to deal with problems arising from bad relationships, divorce, difficulties raising children, impending death, accidents or illness; or just plain feeling empty and directionless—the types of things that drive people into therapy. Because Hermes represents the relationship between consciousness and the unconscious and is the messenger between these two realms, he is symbolically described as inhabiting the boundary, the gap in the wand. Here he is the god of dreams and synchronistic events and the presenter of symptoms that arise from the underworld, the unconscious.

Priests in the ancient Greek healing temples of Asclepius, the Greek god of healing, would interpret patients' dreams received in a dream incubation ritual to determine which god or goddess the diseased person was to follow. This was a central part of the healing process. Lopez-Pedraza sees Hermes as the god that "can connect the patient to the archetype which made him ill, providing healing by this connection." (Lopez-Pedraza 1977, p. 25 note 15) He helps connect the individual, personal life in time/space to the universal, timeless themes, thus giving one a sense of meaning, destiny, and one's place in the universe.

Sioux Holy men interpret visions and Big Dreams received on vision quests, traditionally engaged in by all adolescent males, to discern what is the individual's spirit animal or power. Without a guide from the realm of the spirits a male was unfit even for battle. Vision quests are one of many forms of initiation, each designed to place the initiate in a liminal space (the gap in the wand) between an existing life-style and a hoped for new direction in life.

As mediator between human and animal, Hermes gives voice and music and other forms of expression to the depths (cf. the Silenos-Hermes connection). Greek Corybantic rituals used music to diagnose, heal and discover what god was troubling a patient. (Lopez-Pedraza 1977, p. 25 note 15) Songs usually accompany the visions of shamans, which they use at the beginning of every ceremony to re-connect them to the spirit realm and their particular power.

Hermes is the god most associated with the shaman whose tribal position is to carry on discourse with the realm of the spirits so as to guide and heal the tribe. As with Hermes, the shaman/"medicine man" is the one who knows and promulgates the tribal myths and creation stories. (n 39) The shaman heals not only with medicines, but mostly through re-connecting the individual to the tribal myths wherein meaning is found through knowing one's place in the universe and the proper relationship to the spirits (*Religio* means "linking back").

The wand's gap is also the space associated with the phenomena of Winnicott's transitional object (Appendix E) and the domain of attachment and bonding theory so important in current psychoanalytic theory. One must be cognizant of what level one is working on and not mix levels. One can observe attachment and bonding between particular individuals, then look at the deeper level of the archetypal dynamic being played out in the relationship, and at a still deeper level consider the basic concept of relationship and the implications of the space in Hermes' wand. Because Hermes is the source he is the god that knows and links different levels after setting the boundaries/spaces between the gods by elaborating their genealogy. This establishes their pecking order and domains. He becomes the god of elucidation by delineating the forms and levels through clarity of verbal expression. His most powerful expressive modalities are dreams, myths, music, and other types of imaginal expression, including movement.

The image of the caduceus can also be used to re-format Hermes' relationship with the Goddess and the evolution of consciousness. The lower part of the caduceus could be imagined as being a sphere, a universal symbol of the Self that contains all opposites. This has been called the Tao, the Ground of Being (Tillich), *Inyan* before creation (Lakota Sioux); the original *uroboros* (alchemy), the Great Goddess before calling forth Hermes from Herself, the collective unconscious and the *archetype per se*. Creation begins in some Greek mythologems by the Great Goddess calling forth and then evoking a response from Hermes—he gets an erection, becomes phallic, when catching a glimpse of the Great Goddess. This recognition, reflection and response is the original sin in Judeo-Christian mythology, particularly that substantial element of Christian theology influenced by St. Augustine's interpretation of the Hebrew scriptures.

Hermes is related to the Goddess as the lower sphere on the caduceus by being the "soul-realm as the primordial foundation of all actualizations in life...a middle realm between being and non-being." Hermes as the primordial mediator and messenger "stands on ground that is no ground" and creates a way and conjures up a new creation out of this "trackless world—unrestricted, flowing, ghost like," also associating him with the soul-conjuring wand of the wizard and the necromancer. (Kerenyi 1976, p. 77) This is clearly related to the DST concept known as emergent phenomena coming out of the phase transitions that occur when energy put into a system increases its dimensionality to crucial points (see Appendices A and B).

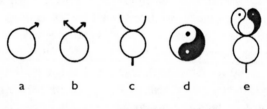

Fig. 3.

Creation occurs with movement associated with the spirit and the "quick" (as in the "quick and the dead"). Hermes is ever active and fleet afoot, and movement going in a particular direction is a vector symbolized by the phallic male sign (fig. 3a). From endless possibilities, one direction is taken, an orientation is begun, time/space is established; creation emerges out of chaos and potential. Consciousness is created when one thing is distinguished from another.

Things are often depicted as a two-ness in dreams when something is ready to cross over into consciousness after gestating in the unconscious. The acceptable element enters consciousness and the other element gets suppressed or repressed because it is not amenable to one's established worldview. This is depicted as a counter arrow on fig. 3b. The repressed elements remain as shadow figures in the unconscious making their presence known in dreams, complexes, symptoms or visions, as in St. John's visions in the Book of Revelation. (see volume 1, chapter 2, section vii)

The experience of something entering consciousness or pressing up from the unconscious has many elements associated with Hermes. It may feel like a great boon or a sudden loss, even both. We can't control it—it happens to us. Vis-à-vis the unconscious material, we feel like we're in the dark, hence Hermes' association with the night: a time of danger, uncertain depth perception, the cover of darkness for thieves and robbers and killers, and nothing standing out clearly until it is right upon us. (Kerenyi 1976, p. 48, 49) The new psychic material may feel threatening and be driven back from sight: it's dangerous to be a baby, so full of potential yet so vulnerable to the reigning psychic powers. This is hexagram 4 in the *I Ching*, Difficulty in the Beginning.

If birth and childhood are survived, consciousness grows and the unique, individual aspects of one's psychic nature are established: the archetypes takes form, are incarnated. Consciousness comes by discrimination, more acutely delineating the differences between things. After attaining their greatest distinctness and separation (fig. 3c.), the opposites can begin to turn towards each other and maintain their distinctness without danger of merging into their opposite. At its fullest development, we have Hermes' wand, figure 2b. The more strongly delineated the opposites and the closer the proximity they be in, the stronger are the energies in the space in the wand and the more powerful, complex and beautiful can be the product that emerges. (n 40)

The *crucial* point is how the opposites interact and how one reacts to something different. There can be antagonism to the point of plundering and attempted destruction as with Hermes and Apollo before Zeus commanded them to be on friendly terms. This led to a clear delineation of their realms and to the ideal relationship between opposites: Hermes and Apollo became best of brother friends at Zeus' command. Hermes took the lead in this process because of his "stealthy," assertive phallic nature used in a friendly and conciliatory yet powerful manner. Union was facilitated by Hermes' knowledge of the mythic roots and stories, his groundedness in the feminine and animal realm, his playful and cleverness and creative nature, and his association with the invention of music. Hermes' reconciliation with Apollo is a mythic story from our Western roots that can help address the rift in our god image that runs like a fractal through individuals and our culture.

A healthy relationship of the opposites can be illustrated by incorporating the Chinese yin-yang symbol (fig. 3d) into Hermes' wand (fig. 3e). The yin-yang symbol shows a complementary nature and a fluid interaction of the opposites, each containing the seed of its complement within. The original image for yin and yang were the two sides of a mountain as the sun passed over it during the course of a day. In the morning one side is lit and the other dark: the situation is reversed in the afternoon. The mountain is the same; the illumination, the perspective, changes. Combining the yin-yang symbol with Hermes' wand illustrates in another manner that each element is of equal value and it takes both to present the total picture, a totality: the parts complement each other to reveal the whole. The masculine has been described as materialized spirit, the feminine as spiritualized matter—two different ways of describing and reaching the same goal depending on where one starts. Wholeness is the union of the two, matter and spirit, matter made conscious, the spirit incarnate in time/space.

The dots within the yin and yang illustrate that they are not unrelated to each other, indeed they exist only because they can be distinguished from their other half ("What is the sound of one hand clapping?" the Zen Buddhists ask). If we think of the dot within as being an eye, what each sees in the other as its opposite or something different is really part of itself at a deeper unconscious and holistic level. "Other" can appear as foreign, strange, mysterious, unknown, repulsive or attractive and its behavior towards us is largely determined by how we interact with it.

Life's, and our planet's, energies will be gobbled up and destroyed if opposites remain bitter antagonists. Zeus' challenge to us is the same he gave to Hermes and Apollo—be on good terms with your opposite. Don't go to the extreme of identifying with it but adopt an attitude of *viva la difference*! Many aspects of the process leading to wholeness and the union of opposites were described in an archetypal numerical manner in a favorite alchemical dictum attributed to the Coptess or Jewess, Maria Prophetissa. (see Appendix J)

To journey well in the gap in Hermes' wand is to be in a state of dynamic receptivity called "beginners mind" in Zen, a virginal attitude from which all wisdom arises. In such a space one can genuinely meet and understand the other: as Thoreau said, really "see" the other and not just look at them. To have a feeling for the other one must have "empathy, appreciation, and enough knowledge to respond appropriately when we meet them." (Lauch 2002, p. 19)

CHAPTER 9

Hermes and Sex

We close our exploration of Hermes' nature where we began—with Hermes' relationship to sexuality. Jung saw sexuality as the central element in the dark side of God, sexuality as God's chthonic roots and the connection to our bodies and to nature. Sexuality is particularly important to males for connecting to their bodies and their natural (nature) side. An alchemical image depicts a naked male lying on his back pointing to a tree growing out of his groin. (CW 12, p. 256) This symbolizes that the movement of psychic energy is particularly associated with sexual, erotic energy at that state of the alchemical process. The comparable image for a woman has a tree growing out of her head. (p. 268)

Sexuality is experienced as a powerful drive in most men; men feel driven by it. It is often seen as shameful and gets repressed, then projected onto women as being evil temptresses. Male sexuality is extremely problematic in Western Christian culture, exemplified by the difficulty of imagining our central Self image, Christ, with an erection (a phallus and not just a penis). We have no sacred images for sexuality; no archetypal presentations of gods and goddesses to frame our sexual natures and put sex into a sacred context. Restricted imagery not only limits Western religion and distorts Western culture, it also affects the core and foundation of psychoanalytic thinking. (Lopez-Pedraza 1977, p. 67, 69) (n 41)

Hermes and Aphrodite have the most associations with sexuality, yet all Greek gods and goddesses have their own type of sexuality befitting their basic characters. (Lopez-Pedraza 1977, p. 66) Approaching sexuality and mythology with this insight allows for the complete range of human sexual activity to be contained within an archetypal matrix and not confined to limited sexual "norms." (p. 67-69)

An illustrative tale is told by Homer in the *Odyssey* about the adultery of Aphrodite. Aphrodite, goddess of love and sensual beauty, was married to the crippled, creative god of the forge, Hephaestus. Aphrodite and Ares, god of war, were having an affair. When the Sun so informed Hephaestus, he forged an invisible chain net, light as gossamer, "a masterful piece of cunning work." Hephaestus strung this about the bed, and trapped the two lovers in the act after he pretended to leave home. Aphrodite and Ares were bound so tightly they could not move a limb to escape. After hearing from the Sun of their capture, lame Hephaestus hurried back home to angrily confront the two lovers, shouting so loudly in his rage that all the gods could hear him. He threatened to keep the lovers "imprisoned on the bed until Zeus returned the gifts Hephaestus had given him in order to win Aphrodite for his wife." (Lopez-Pedraza 1977, p. 60) The gods were attracted to the spectacle, but the goddesses stayed away out of modesty. The gods stood at the door making joking comments and laughing uncontrollably about the clever device containing the naked lovers. Apollo asked Hermes if he would care, "though held in those unyielding shackles, to lie in bed by golden Aphrodite's side?" Hermes replied, "there is nothing I would relish more. Though the chains that kept me prisoner were three times as many, though all you gods and all the goddesses were looking in, yet would I gladly sleep by golden Aphrodite's side." (Homer 1966, p. 131 quoted in Lopez-Pedraza 1976, p. 60)

All laughed except Poseidon, who was not amused. He urged Hephaestus to free Ares from the net. He, Poseidon, would make the required atonement for Ares if Ares would not pay his debt. (Lopez-Pedraza 1976, p. 60) The lovers were then freed.

An imaginative approach to this tale is presented by Rafael Lopez-Pedraza in his seminal book, *Hermes and His Children*:

> Hermes fantasizes sexuality, telling us about a sexuality easily carried and graciously fantasized in the most overt way, to the point of shamelessness. Moreover, Hermes is not at all bothered by telling his fantasies in front of the rest of the Gods. The image shows clearly that Hermes can accept the pornographic play of the psyche, that archetypal fantasizing of erotica which comes to the scenery of our mind via Hermes. This image gives us a spectrum from Hermes, who accepts the fantasy totally, to Poseidon who, in his reaction to the situation, seems totally unable to accept it. This image shows a libidinous attitude on Hermes' part and a strongly repressive one on that of Poseidon. (Lopez-Pedraza 1977, p. 60, 61)

Lopez-Pedraza asks if Poseidon's lack of humor can be attributed to the archetypal background behind an attitude of sexual inhibition that represses sexuality and tries to hide it, puritanically reacting to pornographic images. (Lopez-Pedraza 1977, p. 62) Poseidon's view could characterize "the fathers of modern psychology [who] were scientifically-minded, prudish Jewish men" living in a culture imbued with Victorian Puritanism. During that period sexuality was seen as sick, with perversions arising from sexual trauma. Freud et al tried to release this enchained sexuality and "cure" and control the psyche (p. 62):

> In [Hermes'] acceptance of being seen by the rest of the Gods he catches their fantasies through their projection of their own sexuality onto him. Freud saw this archetypal complex and projected it onto the concrete child, projected it causalistically into actual childhood [his concept of the child as polymorphous and sexually perverse]. (p. 66)

The sexual situation described in the tale shows how the three main characters are caught. Ares and Aphrodite are in chains. Hephaestus, possessed by a "gotcha" mentality of a cuckold, is caught by virtue of being the conceiver of the net. Apollo believes "that with his detached questioning he is not caught" but his very question implies that. His distant, lofty attitude in questioning Hermes might be likened to modern research on sexuality. (Lopez-Pedraza 1977, p. 62, 63) Hermes illustrates one of his traits of never being caught. He totally accepts the fantasy and moves it to the extreme. This is a process hint to analysands and analysts who get caught and stagnated by a sexual image. (p. 63)

The overwhelming amount of pornography in our culture could be seen as the irrepressible, pagan, sexual imagery insisting on being "seen." Lopez-Pedraza asserts,

> [We can assume] that this same amount of pornography that we see in the streets is also within us, in our psyche... These pornographic images are archetypal and...our reaction to their outer appearance [is archetypal]. (Lopez-Pedraza 1977, p. 65, 66)

> Pornography...is an expression of polymorphism...The basic conceptions of psychology came out of paganism, for instance, the Oedipus complex, the polymorphous perverse child, the Hermaphrodite, etc. (p. 68)

> A sexuality without polymorphism does not exist. Mono-
> theism needs polymorphous images from which to image
> its own sexuality and onto which it can project what it
> imagines it is not. To think in terms of "normal" sexual-
> ity has nothing to do with the archetypes of sexualities...
> During Renaissance times the Protestant countries rejected
> the appearance of the pagan Gods within a new syncretism
> in religion, philosophy, and life...These repressed elements,
> neither good nor bad in themselves, have now re-appeared
> in a secular flood of pornography though without the grace
> and art of their Renaissance expression. (p. 69)

Lopez-Pedraza and Hillman in their archetypical psychology approach concentrate on the images presented in dreams, symptoms, etc. to move the psyche. Symbols can generalize and categorize images, removing one from their immediacy. The image can be read as an archetypal picture of life, complete in itself, which is more fruitful than talking in terms of concepts. One must stay with the image. (see Appendix K)

The archetypal approach is to distinguish what gods and goddesses are interacting at the core of one's "personal" life situation. This insight helps with a diagnosis and provides some awareness of what complexities one can expect from being involved in such an archetypal "play." Not knowing the characters and the play can lead to a mis-treatment of the case and wound the soul. Each play will have its peculiar transference issues, its own rhetoric and style of working, its own peculiarities. (Lopez-Pedraza 1977, p. 84, 85)

Lopez-Pedraza challenges therapists to discover which "God or Goddess the patient is a child of" (Lopez-Pedraza 1977, p. 94) noting that many classical writers told "of a God or goddess as one parental side of a mortal's begetting...[which] leaves an imprint and marks the meaning of a whole life." There is a reality of "the conflicts and traumas of the family complex" but psychotherapy has been overburdened by literalizing them, missing "the underlying archetypal genealogy of the patient." A mythic genealogy releases the patient somewhat from the pressures of personal genealogy and opens new or unexpected forms of psychotherapy and life. (p. 93) "It can open the way for psychic movement within the range of the given archetypal possibilities." Connecting the image of a life situation with a God and accepting the God resolves the situation "within a better perspective, more psychological, more human if you will." (p. 94)

The implications of this approach are nowhere more evident than in dealing with sexuality in our lives and in therapy. "Sexuality has [its own] instincts and complexes of archetypes behind it," Lopez-Pedraza notes. "And it is polymorphous since there are different Gods in all of us with their various forms of sexuality which probably express themselves at different times during a lifetime." (Lopez-Pedraza 1977, p. 68) (n 42) Monotheism tends to reduce "proper" sexuality to heterosexual relationships, looking at sexuality in terms of relationship:

> A monotheistic sexuality could be conceived as being an idealized, conceptualized, even messianic sexuality (all that missionarism around the orgasm, the proper orgasm, the proper sexuality in marriage, etc., and projecting onto this a healing function). (p. 67)

> The...sex therapist practices a psychotherapy through pornography under the guise of a clean, scientific, sexual teaching. (p. 65)

Lopez-Pedraza believes the root of the conflict between polytheism and monotheism is sexual, going back to the basic issue of a lack of sacred sexual imagery in monotheism. (Lopez-Pedraza 1977, p. 67) The example mentioned before was how Freud's monotheistic view of sexuality led to the many (polymorphous) forms of sexuality in the pagan gods getting perversely projected upon the innocent psyches of children.

Lopez-Pedraza provides insights and archetypal containers for many difficult sexual situations. He suggests there is an archetypical aspect to our concepts of "virgin nature" which includes a sense of religious fanaticism and a peculiar vulnerability to rape and pillaging. (Lopez-Pedraza 1977, p. 109) (n 43) Hermes fathering of "the moon-like and dark Pan" (Kerenyi 1976, p. 87) reveals the dark side of Hermes nature. The archetypal base of masturbation is Pan, the god who discovered it. (p. 83) In *Hermes and His Children*, Lopez-Pedraza attempts to put an archetypal framework on male homosexuality, panic and nightmares, epilepsy, masturbation, madness, "cheap" sexuality, consort sexuality, rape and masochism, nymphomania, compulsive and manic sex, the Marquis de Sade, Don Juan, the grotesque and freakish in our lives, and Picasso's erotic drawings in old age.

CHAPTER 10

Hermes—God of Ecopsychology and Complexity Theory

Building on my definition of ecopsychology, "the ecology of psychology and the psychology of ecology," we have seen how Hermes can be the god of ecopsychology. First, he is the god of psychology because his traits and activities are associated with our relationship with the unconscious, and it is the unconscious that determines 90% to 95% of our conscious attitudes and behaviors. Jung said, "We need more psychology. We are the source of all coming evil." Of primary importance to Jung was the establishment of a good relationship between the ego and the unconscious as a way of making effective changes in one's life, developing a sense of wholeness and fulfillment through individuation (Jung's antidote to consumerism), and establishing consciousness within a mythic and spiritual framework. *The Homeric Hymn to Hermes* depicted many of the processes and characteristics of the ego-unconscious relationship, revealing essential elements of Jungian psychology and ecopsychology. Hermes takes us to the root of the human psyche and its origins, not only by linking us to our sexual and "animal" sides but also to our nature as mythic beings whose life stories are put into an archetypal, sacred context with Hermes' help. Looking at the psyche through Hermes' eyes enables us to see the varieties of archetypal relationships between humans and ways of linking and mediating between the absolutes of the different religious traditions. To approach ecopsychology from the deepest, most basic levels, we need an ecology of religions and spiritual life. Hermes' caduceus can be a guide in imaging the processes for possible symbiotic relationships between religions. This is important given the contentious religious elements in the post 9-11 world.

Hermes' wand symbolizes the most effective means of dealing with opposites and contending parties at any and all levels: delineate the differences at their basic levels with Hermetic elucidation, then work with Hermetic diplomatic and Eros skills to bring the opposites into a relationship as exemplified by Hermes and Apollo becoming loving brothers instead of civil-warring enemies.

Ecopsychology is not only about nature and the environment: it's about getting our human house in order as well. This includes our social, governmental, institutional, educational, economic, judicial and spiritual systems. It's a given that individuals have to do their own psychological work, and the systems they live in can either support or hinder that process. We are social animals, and "what is done onto the least of us is done to all."

Improving relationships within our own species is not unrelated to a second important dimension associated with Hermes: our human connection to our bodies and sexuality and through our bodies to nature. Lopez-Pedraza identified the major problem in the split in the Western psyche—we have no sacred imagery to carry and archetypically contain the most basic drive in all living things—the sex drive. One of Hermes' main images is a prominent phallus, a sacred image whose absence in Christianity has had a particularly negative effect on Western Christian males with subsequent negative consequences for Western females. Hinduism does a better job with this issue with its *Kama Sutra*, erotic imagery at temple sites, and depictions of a main god (Shiva) as a gigantic phallus. (volume 2, p. 99, note 10) Hermes' response to Ares and Aphrodite being caught in the act of love-making allows a sacred place for erotic fantasying sorely lacking in monotheistic religions, resulting in the plethora of pornography now amplified by the Internet.

Hermes links human to the vegetative realm of the human psyche— the autonomic nervous system that controls sneezing and digestion, heartbeat, etc. We share these deeper levels with the animal kingdom. In one important respect, humans are simply conscious animals; animals that "naturally" use mythic, symbolic and sacred imagery to frame and contain their animal experience. It is Hermes who produces the dreams and visions in those liminal spaces that bridge the gap between the purely physiological "animal" realm and the mind, intellect and spiritual-conscious realm.

Hermes has dominion over the animal realm, "inspirit[ing] the warmest animal darkness," and is also associated with healing plants and the vegetative realm. At the deepest levels, Hermes goes beyond the scientific realm of causality, matter and energy, for he is the god of synchronicity; the acausal, meaningful connections between things that point to another level of reality. It is through these and other associations (like initiations and shamanism) that Hermes is our link to "the indigenous one within," the "two million-year-old man" that Jung said lives within each of us. At the deepest levels of every human psyche we can connect to the land in a sacred manner as "native" people do—that capacity is part of our archetypal, human inheritance. But to do this we need cultural, educational, religious, and psychological systems that recognize and honor these modes of perception and response. When we discover our indigenous nature as Westerners, we will finally be able to appreciate and connect with our Native American brothers and sisters on this continent and begin to learn from their wisdom of the land.

Hermes can take psychology beyond a focus on statistics and rat and monkey experiments and carry us a long way towards healing the Western split from the environment, a split that Jung believed goes to the core of the Western psyche. Jung said we don't have a monotheism, we have an antagonistic split between Jesus and the Devil. An approach to healing that split might be to imagine Hermes and Christ becoming brothers. After all, it was one of Hermes' children, Pan the goat god, who became the template for the Christian image of the Devil.

Hermes is truly the god of complexity theory, the new mathematics that describes the dynamics of much of the universe. Complexity theory reveals patterns within many of the irrational, non-linear processes of life and the inorganic realm and provides the mathematical base for the emergence of new, more complex systems. The dreaming brain, ion transfer across membranes, patterns in volcanic eruptions and stock market fluctuations are prescribed by complexity theory. I think of complexity theory as the mathematical manifestation of the archetypal feminine in the Age of Aquarius that Jung anticipated.

And lest we get too ethereal with these considerations, remember that Hermes is also the god of businessmen. Business and economic systems are a vital part of our environmental dilemma, and will have to play a central role in addressing the problems. The business mentality will have to be known and understood if we are to effectively incorporate it, and our personal understanding and experience of how Hermes works in our own psyches will be of invaluable assistance in that process.

We need courage, imagination and boldness in these challenging times. "Life takes on a special flavor as soon as Hermes becomes your guide" (Jung and Kerenyi, *Le Divin Fripon* quoted in Paris 1990, p. 59) Kerenyi and Jung, true followers of Hermes, believed:

Who ever does not shy away from dangers of the most profound depths and the newest pathways, which Hermes is always prepared to open, may follow and reach, whether as scholar, commentator, or philosopher, a greater find and a more certain possession. For all to whom life is an adventure—whether an adventure of love or of spirit—he is the common guide. (Kerenyi 1976, p. 91)

NOTES

1. Kerenyi summarizes the world of Hermes. In it he sees:

> The sum total of pathways as Hermes' playground; the accidental "falling into your lap" as the Hermetic material; its transformation through finding—thieving—the hermetic event—into an Hermetic work of art, which is also always something of a tricky optical illusion, into wealth, love, poetry, and every sort of evasion from the restrictions and confinement imposed by laws, circumstances, destinies. (Kerenyi 1976, p. 54)

> "What occurs in [Hermes' world] comes as though from heaven and entails no obligations; what is done in it is a virtuoso performance, where enjoyment is without responsibility. Whoever wants this world of winning gains and the favor of its God Hermes must also accept losing; the one is never without the other." Hermes is therefore "the spirit of a form of existence which under various conditions always reappears and knows both gain and loss, but shows kindness and takes pleasure in misfortune. Though much of this must appear questionable from a moral point of view, it is nevertheless a form of being which with its questionable aspects belongs to the basic images of living reality and therefore, in the Greek view, demands reverence, if not for all of its various expression still for the totality of its meaning and being." (Walter F. Otto, 1954, *The Homeric Gods*, Moses Hadas, trans., Pantheon: NY quoted in Kerenyi 1976, p. 3-4 with translations from the German original by Murray Stein)

2. The Greek myths and rituals were set in pre-time as are Native American and other indigenous cultural myths. I will use primarily *The Homeric Hymns* translated by Charles Boer (1970), attributed to have made the "Greek gods appear not as abstract presences, but as moving and radiant eruptions of the sacred." (William Arrowsmith quoted on back cover of *The Homeric Hymns*)

3. The Greeks believed the universe created the gods and not vice versa: Heaven and Earth sprang from Chaos and became the parents of the Elder Gods, the Titans. The Titans were gigantic figures of incredible strength that reigned for untold ages. Some of their children became Olympic gods. The Titans' powers were transferred to the Olympians or the children of the Olympic founders. Cronus (Saturn) ruled over the Titans and his wife Rhea (Ops) bore Zeus who later dethroned his father Cronus (Hamilton 1942, p. 21, 22):

The other notable Titans were OCEAN, the river that was supposed to encircle the earth; his wife TETHYS; HYPERION, the father of the sun, the moon and dawn; MNEMOSYNE, which means memory; THEMIS, usually translated as Justice; and IAPETUS, important because of his sons, ATLAS, who bore the world on his shoulders, and PROMETHEUS, who was the savior of mankind. These alone among the older gods were not banished with the coming of Zeus, but they took a lower place. (p. 22)

Maia's ancestry is definitely pre-Olympian. Her father was the Titan Atlas and her association with the sky as the eldest of the Pleiades and her epithet "majestic" associate her with the Titan period. (Kerenyi 1976, p. 19)

4. In the *I Ching*, the archetypal feminine in the Zeus-Maia relationship is that of the virgin as depicted in hexagram 44. See Appendix G: the Sacred Prostitute and the Erotic Feminine.

Hermes being worshipped as a phallus or with emphasis on his phallic nature underscores the dimension or level of the psyche he is associated with. Whereas every god and goddess has their own unique expression of sexuality, Hermes brings a focus on sexuality per se, particularly archetypal masculine sexuality and that *in relation* to archetypal feminine sexuality. Even more basic is the "accidental 'falling into your lap'" aspect of Hermes that seems to come through from heaven without obligation and "where enjoyment is without responsibility" (note 1), an aspect of Hermes that Kerenyi sees as a "residue of the chaotic, primordial condition." (Kerenyi 1976, p. 24) This places Hermes' energy as that close to the source of the psyche, in the space between the Tao and its first basic manifestation—yin and yang. The additional emphasis on being conscious of this dimension or level of psychic experience can be associated with such Buddhist concepts as "thoughts without a thinker," the pregnant void, and being above/below the gods as that which gives rise to the gods as Jung thought of Buddha's goal. (Jung 1961, p. 279)

5. The archetypal "feminine" activity of nurturing, containing and bringing inspiration (hexagram 1) into space-time as form is associated in the *I Ching* with hexagram 2: The Receptive. This is Sophia in the Bible who puts God's thoughts into material form—the universe. (see volume 1, chapter 2 note 37)

6. The turtle or tortoise has worldwide mythological feminine associations. Its bulk, obstinate strength and stubby supportive legs feed into a mythic imagination as the pillars, stabilizers or bearers of the cosmos. These elements in association with earth or water and the tortoise's longevity (immortality) lead to the ideal of fertility, of primordial waters that are ruled by the Moon. Primeval waters are associated with the fertility of generation and regeneration. Being close to the earth with its short

legs lends further associations to mother earth, as does steadfastness, slow but steady progress, and a material evolution that is in contrast to the brilliant changes of the spirit. (Chevalier and Gheerbrant 1994, p. 1016, 1017)

The tortoise is also associated with Cancer in the Zodiac, "symbolizing (involutive) chaos, with the hope of renewal of life." (DeVries 1974, p. 471) It is a symbol for the raw material, the *prima materia*, for the start of the alchemical process—the beginning of the spiritualization of matter. (Chevalier and Gheerbrant 1994, p. 1018) Hermes creation of the lyre for making music after killing the tortoise is an example of this process.

The tortoise is a symbol of the universe because its shell is representative of the half curved dome of heaven and the flat surface of earth. (Chevalier and Gheerbrant 1994, p. 1016) In this sense it is related to the collective unconscious, "the realm of the mothers," the matrix containing all opposites out of which everything emerges and into which everything is resorbed.

The opposites are indicated by the perceived androgyne nature of the tortoise: the shell has been associated with the vagina and the emerging head as a penis that can protrude or be withdrawn. (Chevalier and Gheerbrant 1994, p. 1019)

Because the tortoise lives between the symbolic heaven of its dome and the earth of its flat base, it is thought of as being a mediator between these two archetypal opposites—a Hermetic position. "For this reason the tortoise was adjudged to possess the powers of knowledge and prophesy" and was used for fortune telling in China in conjunction with the evolution of the *I Ching*. "It must be due to its virtues as the omniscient and beneficent ancestor that the tortoise is so often the companion and family friend of human beings," Chevalier and Gheerbrant note. (Chevalier and Gheerbrant 1994, p. 1018) Hermes was also known as a god friendly to man.

7. Eros was the grandson of Craft and the son of Resource (Poros) who lay with Need (Penia) who thought "to get a child by Resource would mitigate her penury." (Kerenyi 1976, p. 55) It was the fate of Eros to always be needy, "But...he brings his father's resourcefulness to his designs upon the beautiful and good, for he is gallant, impetuous and energetic, a mighty hunter, and a master of device and artifice—at once desirous and full of wisdom, a lifelong seeker after truth, an adept in sorcery, enchantment, and seduction." (Plato, *Symposium*, in *The Collected Dialogues*, edited by Edith Hamilton and Huntington Cairns, Bollingen Series LXXI, New York, 1961, p. 555, 556 quoted in Kerenyi 1976, p. 56)

8. Myths evolve as cultures evolve. (Brown 1969, p. 44-46) By Homer's time (1500-1200 BCE) Greek culture was dominated by kings who had absorbed the autonomous familial collectivities into larger economic and social units. The newly emerged social classes were arrayed into

"a pyramidal structure subordinate to the kings." (p. 49) "When social differentiation had arrived at the point of class divisions, the landed aristocracy secured a strangle hold on the instrumentalities of state organization." (p. 48) Trade and industry were concentrated around the palace reflecting the dependence of the craftsman upon the king. (p. 51) The relevant myth was of Zeus as the monarch of the Olympic pantheon "and the component gods were given ranks and positions analogous to the component orders of society." Hermes became subordinate to Zeus as his messenger and servant-in-chief, having lost his status "as an independent and autonomous trickster." (p. 49, 50) He was god of "a class of professional boundary-crossers"—Hermes was associated with the pioneers who went into the wastelands to look for raw materials (and discovered iron-making), the merchants who crossed boundaries to trade, and the skilled craftsmen who wandered about to sell their labors to the highest bidder. (p. 46) The economy of the royal households depended on the skilled and unskilled workmen (p. 51) and hired labor. (p. 52) The riches these activities brought to archaic and classical Greek society made Hermes a cultural hero. (p. 38)

9. Norman O. Brown notes that a sophisticated audience, probably that of the tyrant Hipparchus (note 20), is implied by the frequent poetic parodies of Homer and Hesiod. (Brown 1969, p. 134) "The verbal humor is for the most part based on a calculated incongruity between the subject matter and the epic language used. For example, the formula 'craving meat,' applied in the *Iliad* to a ravenous lion, is used to describe the newborn baby setting out after Apollo's cattle. The whole scene of Hermes and the tortoise is...[a] classic parody of the Hesiodic maxim, 'It is better to stay at home, since the outside world is noxious.'" (p. 134) Hermes told the tortoise, "You'll find it much better at our house— outside here things are bad." (Boer 1970, p. 20) "[Hermes] makes a mocking allusion to the traditional and rustic use of the tortoise as a charm—'While you live you will be a good charm, if you die you will become a pretty singer'—and then he proceeds to kill her." (Brown 1969, p. 80)

10. Creativity is associated with the transition phase in dynamic systems theory, a mathematical dynamics with Hermes as its god. (see Appendix E: Hermes as God of Dynamic Systems Theory) Hermes is related more to the virginal creative aspect of the feminine rather than the containing, nurturing mother aspect. (see Appendix G: The Sacred Prostitute and the Erotic Feminine) Myrtle is Aphrodite's plant along with the rose. This combination of plants is associated with the emotional nature of the love goddesses who suffer the loss and death of their beloved who is symbolic of forms that emerge from the archetypal feminine matrix only to be deconstructed and have their energies merge back into the earth (unconscious matrix). (see Appendix G)

11. Hermes is the god of activities requiring skill and dexterity, like wrestling and gymnastic exercises. (Bullfinch 1962, p. 37) His power would be in relation to his grandfather Atlas who holds up the world.

    Skill and dexterity are also linked to Hermes the Craftsman, making Hermes a cultural hero. Hermes the Craftsman originates with primitive magic as an attempt to manipulate the world. "The Greeks attributed the discovery of iron-smelting to magicians (the Idaean Dactyls)," Brown notes. (Brown 1969, p. 21 note 32) Magical practices supplemented the techniques of primitive craftsmen "and success at his craft is taken to indicate possession of magical powers," (p. 21) for without knowledge of causation the transformations wrought by metallurgy looked like magic:

    > In the *Odyssey* a journeyman is said to owe his proficiency to Hermes, who "bestows joy and glory on the works of all mankind." As a craftsman-god, Hermes is endowed with the essential traits of the mythological type of culture hero, of which there is no finer example than the Greek Prometheus. Like Prometheus, Hermes is represented as "pre-eminently intelligent": the primitive mind knows not our antithesis of mental and manual labor and regards craftsmanship and mental ability as going hand in hand. Like Prometheus again, Hermes is represented as a friend of mankind, a source of material blessings, "the giver of good things," "the giver of joy." (p. 21, 22)

    Hermes in the *Hymn* is "the little Prometheus" who releases Promethean energy from the shackles created in Hesiod's myth. (p. 89)

    The Greek word for "trick" in Homer's time had connotations of magical action and was "also used interchangeably with the usual word for 'technical skill' and the 'gain' and profit achieved with its products." (Brown 1969, p. 22) This is "the good" which Hermes gives and Hermes the trickster as cultural hero. (p. 23, 24) Trickery as magic skill appears in the *Hymn* with Hermes sandals being described as the work of a mighty demon and the lyre represented as a miraculous creation. (p. 82) The Homeric Greek for "steal" has connotations of magical trickery and denotes technical proficiency; "stealthy" as the root for technical skill "derives from its basic meaning of "secret mysterious action." (p. 23)

12. Norman O. Brown notes several aspects of Hermes sacrifice to the gods in the *Hynm* that help pinpoint the time and place for the origin of the *Hymn* at 519-520 BCE Athens (Brown 1969, p. 117-127), namely Hermes use of a firepit instead of an altar to sacrifice to the twelve Olympian gods (p. 119-124, 126-127), not eating any of the sacrifice himself (an association to Hestia worship where no one received a share of the sacrifice (p. 125) and Hestia's connection with the Olympians (p. 124-126), and the use of sticks to start the fire as opposed to the

lens used in the Acropolis. (p. 126) Brown suggests the firepit would have been in the agora (market place), the center of activity of the merchants and craftsmen. These features point to a particular moment in the evolution of the agora as a new center of the political power for the merchants and tradesmen in their struggle for equality with the landed aristocracy. Their sacred center was in the Acropolis. (see note 20)

13. Hesiod turned Hermes into the Greek equivalent of the serpent in the Garden of Eden with his myth of Pandora. It was his reaction to the evolving Greek culture and economy. The city agora developed into a market place during Hesiod's time (the seventh century BCE) and the newly emerged social class of merchants and craftsmen "were now successfully emancipating themselves from their previous dependent status." (Brown 1969, p. 62) "The new ethic of acquisitive individualism conflicted with the traditional morality which the Greeks called Themis—the body of customs and laws inherited from the age of familial collectivism." Brown describes Hesiod as "the first nostalgic reactionary in Western civilization...[He] recommends agriculture as the best way of life because it offers the maximum self-sufficiency, the maximum isolation from the new economy...Hesiod is an isolationist: 'it is better to stay at home, since the outside world is noxious'; he firmly turns his back on the new commercial culture." (p. 63)

The cultural change is reflected in an epithet for Hermes that originally meant "tricky" that came to mean "good for securing profit." "Hesiod uses 'trickery' as well as 'robbery' as abusive terms to describe the ethic of acquisitive individualism." He talks of the shamelessness of wealth attained through "plunder through his tongue, as often happens when gain...deceives men's senses." (Brown 1969, p. 64)

Hermes is the god "friendliest to man" in Homer because Hermes' "stealthy disposition" is always used for benevolent ends. In Hesiod's myth of Pandora, Hermes is given a satanic aspect. Zeus called for the creation of Pandora to punish men for Prometheus' theft of fire from the gods:

[He had] Hephaestus fashion a woman out of clay, Athena equip her with handicraft skill, Aphrodite with beauty, and Hermes with "the mind of a cur and a stealthy disposition." Zeus calls her Pandora because all the gods have endowed her with gifts. Pandora is conducted by Hermes to Epimetheus, who accepts her, against the advice of his brother Prometheus. Then she takes the lid off the jar in which her gifts are contained. Whereas men have lived up to this time free from ills or toil or sickness, now "the earth is full of ills, and the sea is also full of them," and only Hope remains in the jar. (Brown 1969, p. 54)

Brown comments on Hermes "special responsibility for the catastrophe. Pandora opened the jar with malice aforethought: "she had in her mind bitter sorrows for men"; and her maliciousness—"the mind of a cur and a stealthy disposition"—was bestowed on her by Hermes," putting Hermes in the role of the serpent who tempted Eve. (Brown 1969, p. 54, 55)

Originally Pandora symbolized the bounty of the earth and Prometheus (or Epimetheus) represented the fertility-promoting action of human labor and skill—agriculture—to reap the bounty of the earth. In Hesiod's myth "[Pandora] is an artifact, manufactured by the craftsman god Hephaestus; she is given to mankind along with Promethean fire, which is the symbol of metallurgy; the gods involved in her creation include the leading artificers and culture heroes of the Greek Pantheon (Hephaestus, Athena, Hermes) and the leading artificer hero, Prometheus. Pandora in Hesiod is the symbol of handicraft culture." (Brown 1969, p. 61)

Hesiod's changes of Pandora's nature have been ascribed to his own misogynism, "for the concept of Earth as the mother of the human race Hesiod substituted the concept of the evil artificial woman as the prototype of womankind...In terms of an agricultural economy, the Earth is considered as (1) the Giver of all, and (2) the Mother of all; in terms of a handicraft economy, (1) the Giver of all (culture) is identified with craftsmanship, and (2) human beings are regarded as artifacts in origin." (Brown 1969, p. 61, 62 note 9) The "stealthy disposition" Hermes gave Pandora "refers to the bewitching guiles of sex appeal... In so far as Pandora...is...a personification of the female sex, Hesiod's negative attitude toward Hermes, the god of sexual appeal and marriage, may be attributed to his profound misogynism." (p. 55, 56)

Brown sees Pandora as "more than a personification of the female sex"; Hesiod mainly uses the myth to present his moral and social philosophy. He is addressing the necessity for work in Pandora as "Zeus's revenge on mankind for Prometheus' theft of fire: 'for the gods have hidden away from men the means of existence; otherwise you would easily do enough work in a day to supply you for a full year even without working'...The opening of Pandora's jar has the same consequences as the eating of the forbidden fruit in the Garden of Eden." (Brown 1969, p. 56) For Hesiod the Golden Age was when Cronus was king of the gods, while after Pandora Zeus and the Olympians determine men's manner of making a livelihood. (p. 57) Hesiod's diatribes on work are part of a civil war with his brother Perses who he "typed as a man who hangs around the agora and prefers to make money dishonestly, particularly by legal chicanery." (p. 81) "Hesiod's misogynism—an attitude which differentiates his from the Homeric outlook...stems from the difficulty he has faced, as a poor farmer, in supporting a wife." (p. 56 note 4) "Hesiod's doctrine is that the culture of his own times is a curse rather than a blessing." (p. 60)

14. In complexity theory terms, when the complex becomes "deathly" unbearable, thereby increasing the energy in the psychic system, a "symbolic avalanche" may occur and the transition phase moves one from the personal-concrete to the archetypal-transpersonal-"spiritual" dimension. (see Appendices A and B) Hermes' creation of fire by rubbing two sticks together to use in the volatilizing sacrifice of the cow meat is metaphorically depicted by the close jostling of the elements of the conflict that generate energy out of which may (or may not: Hermes) emerge light, insight and meaning.

15. In the Grimm's fairytale "Iron Hans" (titled "Iron John" by Robert Bly) the boy's theft of the key from under his mother's pillow is a hermetic theft. This key unlocks the Wildman's cage and begins the boy's initiation. Hermes is the god of binding and unbinding which is related to him being the god of initiations. My thanks to David McKee for pointing out this connection in "Iron Hans."

   Hermes' attributes as the trickster/magician in archaic Greek society were stealthy and secret activity, associated with mystery. The archaic Greeks organized at the level of clans and tribes did not distinguish between trickery and magic. (Brown 1969, p. 18-21) Hermes' was a type of thievery not associated with the brute force of robbery—that was linked to heroic actions such as Hercules being "the typical cattle-raider of Greek mythology." (p. 5-7) Hermes theft of Apollo's cattle emphasized stealth and "not the usual resort to open force." "Habitual stealing produces the cunning trickster, habitual robbery the fighting hero," writes Brown. (p. 7)

16. The nine muses, described as the lesser gods of Olympus, were a band of lovely sisters who were daughters of Zeus and Mnemosyne. In later times each muse came to have her own special field: Clio (history), Urania (astronomy), Melpomene (tragedy), Thalia (comedy), Terpsichore (dance), Calliope (epic poetry), Erato (love poetry), Polyhymnia (songs to gods) and Euterpe (lyric poetry). (Hamilton 1942, p. 39, 40) The nine linen cords of Hermes' lyre were in honor of the nine Muses. (Bullfinch 1962, p. 37)

   The Muses were companions with the Graces, and together "they were queens of song, and no banquet without them could please." The Graces were "always together, a triple incarnation of grace and beauty." "Agalia (Splendor), Euphrosyne (Mirth) and Thalia (Good Cheer)...were the daughters of Zeus and Eurynome, a child of the Titan, Ocean...The gods delighted in them when they danced enchantingly to Apollo's lyre, and the man they visited was happy. They 'give life its bloom.'" (Hamilton 1942, p. 39)

17. Jung said the psyche cannot get outside itself to be able to take an objective position. "[The psyche] cannot set up any absolute truths, for its own polarity determines the relativity of its statements." (Jung 1961, p. 350, 351)

18. Jung was critical of Freud for missing the symbolic dimension of sexuality and treating it in a reductive manner. The "Emma" dream illustrates the anima at its deepest level as a function of the Self and intercourse as a symbol of inspiration from the Source.

19. Norman O. Brown emphasizes that Hermes and Apollo agree "to *share* both lyre and cattle. Each initiates the other into his own art." (emphasis added, Brown 1969, p. 91) Hermes says let *us* herd the cattle, re-uniting the cows he stole with Apollo's bull and reproducing. (p. 92) Brown interprets this as the lower classes achieving equality and a comment on the relationship between the two cults; "'friendship' may only mean that open hostilities have been concluded." (p. 93) See note 20.

20. Norman O. Brown makes a convincing case that the *Homeric Hymn to Hermes* was written in 519-520 BCE in Athens as an important mythological support for the Hermes' cult in competition with the Apollonian cult. The rising economic power of the merchants and craftsmen for whom Hermes was god were wrestling political power and prestige from the oligarchic aristocracy who honored Apollo. The *Hymn* is rather unique in its assertion that Hermes tussled with Apollo for equality, invented the lyre and taught Apollo how to play it, and discovered music—all these things as a precocious upstart baby! (Brown 1969, p. 93-100) Hermes as a baby "accentuates the contrast between Power and Helplessness, between the established authority of the aristocracy and the native intelligence of the rising lower classes." It symbolizes the birth of a new world and the meteoric rise of the lower classes in Athens. (p. 90)

Plato and others disparaged the acquisitive individualism of the mercantile class, derided a self-interest they perceived as being placed above society and the state, called them liars, accused them of living a "life of desire," and equated them with thieves. (Brown 1969, p. 83, 84, 88, 89) Plato's ethical doctrine was "that all professions in which the end is profit are vulgar and incompatible with the pursuit of virtue." (p. 83) The landed aristocratic Athenians in the last half of the sixth century BCE "identif[ied] trade with cheating, and the pursuit of profit with theft." (p. 84) They looked down upon the often unscrupulous behavior of the local merchants while feeling threatened by their rising economic power.

Brown traces the merchant-thief association to Hermes roots in Greek archaic culture. The autonomous clans and tribes held market festivals around sacred sites at the boundaries, wastelands, or neutral grounds where several villages converged. (Brown 1969, p. 39, 40) "Primitive trade on the boundary was deeply impregnated with magical notions," Brown noted. (p. 40) "Silent" trade was the most primitive form of trade where the seller left, for example, a lump of unworked iron and a remuneration at a boundary spot sacred to Hermes and returned to find it reworked into a useful object. "The object so mysteriously acquired

is regarded as the gift of a supernatural being who inhabits the place."
(p. 40, 41) In another form of primitive trade a man could take possession of an object in full view of its owner, but only if the owner could
reciprocate with something belonging to the other person—a variation
of gift exchange. (p. 42) "Magic is needed to overcome the distrust
of the stranger and break down the taboos on social intercourse. The
exchange of goods affected by the general license to steal could only
take place as a ritual sanctioned by the god of the boundary-stone."
(p. 43) Brides were also "stolen" in this manner. "'Stealing' a strange
woman was a magical act consummated in the rituals on the boundary. Thus Hermes came to be the master of magic art of seduction and
a patron god of marriage." The ritual of the athletic contests at the
intercommunity festivals on the boundary was another occasion for
trade and Hermes was one of its patron gods. (p. 44)

In the third quarter of the sixth century BCE the political power
was moving from the patriarchal aristocracy, whose worship center was
the Acropolis, to the agora, the emerging political and public market
center. *Hermes agoria* and the cult of the Olympic gods as a symbol of
Greek unity sanctified this new locale and the political equality of the
lower mercantile classes at the expense of the aristocracy. (Brown 1969,
p. 114-117) Brown claims "we must assume that there was a struggle to
win for Hermes official recognition as a god of the agora" and it paral-
leled his intrusion into the world of culture. (p. 113)

The tyrant Hipparchus in about 520-514 BCE set up the first herms
to serve as milestones along the country roads of Attica thereby integrating the cult of Hermes into the urban and political life of the city
state. (Brown 1969, p. 112, 113) The first statues of Hermes *Agoraios* was
set up in Athens twenty years later, its characteristic form being a herm
which was an Athenian invention. (p. 113, 114)

> Hipparchus installed Hermes as the guardian of the attic
> road system... motivated by hostility to Delphi [Apollo's
> oracle]; it was clear that he chose Hermes to perform a
> function for which Apollo was equally qualified—Apollo
> was also god of ways, of boundaries, and of the agora, and
> at a later period there were Apollo herms. (p. 115)

In establishing the place and time of writing the *Hymn* (see note
12) Brown also correlates the unique claims in the *Hymn* of Hermes
being the god of music and consort of Mnemosyne with the promotion
of culture in the lower classes. (Brown 1969, p. 127-129) The tyrant
Hipparchus was a strong supporter of the underprivileged, industrial,
and mercantile classes and several stylistic aspects of the *Hymn* suggest
it was written for a private banquet performance for him and his coun-
cil. (p. 131, 132) Aristotle described Hipparchus character as "gay,"
"amorous," and "devoted to literature and music." (p. 134) The sophisticated humor of the *Hymn* "exploits traditional mythological concepts

for comic effect" (p. 132) and the subtle risqué humor includes Hermes description of the lyre as a lovely, shrill-voiced female companion to be enjoyed day or night at lively gatherings. (p. 133, 134)

The poetic parodies of Homer and Hesiod (note 9) and the elaborate descriptions of the construction of the lyre suggests the poet's appeal to a sophisticated audience. The lyre was brought into Hipparchus' literary circle with the introduction of the new Ionian commercial culture. Apollo was associated with the mainland aristocratic culture and an "earnest and didactic" flute elegy. Ionia developed a "more personal and more sophisticated style of poetry...[with] various melic meters" and lyre accompaniment. (Brown 1969, p. 135)

Hipparchus was in the line of Peisistratid rulers who promoted the rising hopes and aspirations of the rising craft and mercantile classes by construction of an enormous gymnasium open to the public. The Perisistratids are associated with "the termination of the aristocratic monopoly of cultural pursuits and the establishment of institutions that made literature, music, and athletics available to the lower classes...[accounting] for the phenomenally high cultural level of Athenian people as a whole in the fifth century." (Brown 1969, p. 128) "Hermes the god of education and culture is the religious symbol of the aspirations and achievements of the Greek lower classes." Accordingly Hermes underwent a transformation from a rustic figure of his origins to being, like Apollo, "a beautiful son of Zeus" ; the self-styled "fair and good." (p. 100, 101) The manner in which Hermes was portrayed in art began to change about the time he began to be recognized as a musical god:

> In early archaic art Hermes is a bearded, muscular, and rather comical figure—a stylized picture of a man who must work for a living. In the sixth century Hermes begins to lose his beard, and becomes, as Apollo had been before him, the image of the perfect young gentleman, the ideal ephebe, the flower of physical and mental culture, refined by the leisure arts of music and gymnastic—the concept immortalized in the Hermes of Praxiteles. (p. 100)

21. Although Hermes primitive associations as trickster-magician survive in the Hymn in such acts as transforming into a mist to re-enter his mother's home and in turning Apollo's tethers into bindings of the cattle, the emphasis in the Hymn is on Hermes the Thief. (Brown 1969, p. 74, 75) (see note 20) This emphasis demonstrates the evolution of the Hermes myth with the economic, cultural and thought forms in Greek society in the Homeric age between 1500 and 500 BCE. (p. v) The Hymn offers a novel presentation of Hermes as an ambitious aggressive upstart vying with Apollo that "is in sharp contrast with Homer's picture of Hermes the loyal subordinate of Zeus." (p. 87) Hermes as god of the agora and craftsmen reflects the ethical principles of acquisitive

individualism associated with a businessman's mentality—"the duty of self-help and the doctrine that money is the man." (p. 83) Hermes dismisses as being childish his mother's scruples about his thievery and says "I will take up whatever business is most profitable" (p. 75) as he seeks a life of luxury and affluence. (p. 80) He seeks equality with Apollo by illegal means if necessary. (p. 80) Hermes frenetic activity and delight in technique are modeled on the individual and original genius of the inventor and the bustle of the craftsmen. (p. 78, 79) Hermes moral philosophy inspires his inventions: he tells the tortoise he is about to kill "I will be the first to profit from you." (p. 80) His manners and morals are on the vulgar side. He demonstrates an agora mentality by being litigious, "skillful at making the worse appear the better reason," lying brazenly to Apollo, and successfully using "a mixture of trickery, bluffing, flattery and cajoling to persuade Apollo to let him keep the cattle" even after Zeus had commanded him to give them back. "These are the essential traits of the impudent and smooth-talking self-seeker that haunted the Athenian agora," Brown notes. (p. 80, 81)

With the merchant being condemned as a thief no matter what he does, "it is only natural for him to react by justifying and idealizing theft." (Brown 1969, p. 85, 86) The *Hymn* ignores the distinction as did Hesiod, Solon and Plato, "between forcible and fraudulent appropriation...by attributing a cattle-raid to Hermes the Thief, and by describing him as a 'robber' and 'plunderer.'" (p. 84, 85) Kings were no longer leaders of marauding bands (p. 3-6) by Hesiod's time. "They now formed a landed aristocracy which had a vested interest in the suppression of all attacks on property, including both robbery and theft." (p. 85)

22. Norman O. Brown traces Hermes' diplomatic and messenger role back to the archetype of the trickster-magician in archaic Greek society where Hermes main attributes were stealth and secret activity. (Brown 1969, p. 18-21) The powers of the magician were primitive attempts to control the dangers in the natural environment through ceremony, songs, chants and/or whispered phrases in ceremonies—Hermes heraldic origins (p. 15, 16, 20, 26-32) and Hermes as the god of eloquence (p. 15 note 21) and song. The rudimentary stage of political institutions "needed the support of religious sanctions, and were organized as religious ceremonies; hence a role was allocated to the ceremonial expert, the herald." (p. 27) With the emergence of a ruling class and kingships in the Homeric-Mycenaean age (1500-1200 BCE) many of the traditions of the magic art were converted "into religious sanctions to bolster their own authority." (p. 29) "The herald is the ceremonial expert in the rituals that center around the royal palace, the public assembly place, and the like. His 'town-crying' function is derived from his ceremonial function" (p. 27):

> The personal service which the herald rendered the king
> was a by-product of this ceremonial functions. In Homeric
> society public religion was the responsibility of the king,

with the result that the ceremonial expert became an acolyte to the king. The king presides over the assembly, the heralds keep order; the king makes a sacrifice, the herald prepares the sacrificial animal; the king has a ceremonial banquet, the herald takes care of such ceremonial niceties as the proper division of the meat into portions. This ceremonial function is naturally extended to include the purely secular [as in diplomacy and message-bearing]. (p. 27, 28)

23. Princeton professor James Gould claims that the bee dance is definitely a language, able to describe "something removed in time and space." (Lauch 2002, p. 140)

24. Norman O. Brown is among the scholars who maintain that the end of the *Hymn*, beginning with Apollo expressing fear over losing the lyre and his bow, was constructed by a different author (argued in Brown 1969, p. 148-155) "The dual authorship is revealed by a clear break in dramatic continuity, inconsistencies and duplications in the narrative, and marked stylistic differences." (p. 103) The revisionist author placed Apollo in a more exalted light, perhaps reflecting Plato and Pindar in their disapproval of Hermes thievery, his claim of inventing the lyre, and for outwitting Apollo. Hermes popularity was disproportionately greater than official recognition. (p. 102 note 56) The revised ending disavows Hermes equality with Apollo, giving Hermes an inferior stature. Apollo gives Hermes the magic wand, Hermes aboriginal possession, and declared "Hermes will fulfill all ordinances which he, Apollo, pronounces in his capacity as the mouthpiece of Zeus." (p. 103) Hermes had not asked for a prophetic device, but Apollo expounds on "the difficulty and responsibilities of his own oracular profession" before giving Hermes an oracle that "does not reveal the will of Zeus and hence is wholly unreliable." (p. 103, 104) The oligarchic concept was that Apollo was the "prophet of Zeus" with his Delphic Oracle at "the navel of the world." Representatives of the Oracle in the Greek cities were the experts on sacred matters. Their objection to mantic devices was that anyone could use them; no skill was required. "[Hermes] was the patron of lottery—of which the mantic dice are one species—and lottery was one of the characteristic institutions of Greek democracy; the extensive use of lottery in the selection of Athenian public officials was the supreme expression of the democratic principle of absolute equality of all citizens." (p. 105)

Apollo gave Hermes protection of an array of animals, "thrusting Hermes down to the status of a purely rustic cult." The fact that Hermes did not object as he would have during "his restless pursuit of power" in the first part of the myth indicated to Brown that Hermes had "been stilled by the characteristically Delphic virtue of self control" with the limitation of desires from knowing yourself and your place. (Brown 1969, p. 104) "Apollo explicitly identifies commerce with theft "when he expresses fear that Hermes will steal his lyre and bow because he "received from Zeus the office of establishing the practice of commerce among mankind." (p. 83) Apollo's hostile judgment closes the myth when Hermes is denied the "right to the epithet 'giver of good'—Hermes 'does little good, but spends his whole time cheating the human race.'" (p. 104)

Brown notes a tendency of mythographers "to reduce the dynamics contradictions of Greek mythology in its vital period to a dull flat consistency" using "tortuous efforts to harmonize." (Brown 1969, p. 97) Both Apollo and Hermes "were gods of music and of divination; both were patrons of youth, and hence of athletics, both were guardians of the house and of roads; both were patrons of pastoral life." (p. 93) Brown sees a Hermes bias in the writer of the *Hymn* particularly with regard to music. (p. 99) Hermes claims to be as competent or even superior in every form of music and he is *the* patron of the tortoise-shell lyre. (p. 94-96) The poet also makes the unique claim "that Hermes, not Apollo, is the companion of the Muses" as the consort of Mnemosyne. Earlier tradition made Apollo the father of three of Mnemosyne's daughters. (p. 96)

25. Kerenyi's work with the Greek text suggests that a formal preparation and ordination are implied by Apollo's "appointment" of Hermes as messenger and escort to Hades:

> Hermes became messenger and escort to Hades only after a preliminary ceremony had been completed...It belonged to the essence of the Greek mysteries that through their initiations one comes into friendly relation with Hades. (Kerenyi 1976, p. 43)

"The God of the mystery is himself generally the first to be initiated." (p. 75) Death is viewed as "not the worst gift" for one initiated into the mysteries. (p. 43)

26. The archetype of the masculine symbolized by upright, phallic shaped stones in association with a spring as archetypal feminine source is found in the placement of the manitou stones by the ancient Cheyenne in Wisconsin. (Herman Bender, personal communication) These placements are over 2500 years old, putting them roughly in the same time frame as the herms-and-spring associations in ancient Greece.

27. In the ancient cult of the nymphs in Attica, "Hermes is expressly assigned to the Goddesses as their permanent escort." In countless votive reliefs from the hills and caves of Attica, he is always shown leading a threesome of nymphs who represent primal feminine fruitfulness that "allow(s) everything to burst into life in the deeps of the caves, the springs, the roots, the hills." With them Hermes is the masculine component of the feminine aspect of untamed fruitfulness. (Kerenyi 1976, p. 60, 61)

The feminine nymphs came into being with the trees and passed
away with them (Hermes' mother was a nymph). They were wet-nurses
for divine and semi-divine children and enjoy the "immortal foods of
the Gods." (Kerenyi 1976, p. 60) The three nymphs relate to the three
aspects of the one Great Goddess; maiden, mother and crone (recall the
three virgin sisters associated with the bee oracle). The original understanding of the triad was a oneness of "a maidenly being, maidenly not
like human brides but like springs and all primal waters, who become
a primal mother and then re-appeared once again in her bride-like,
maidenly daughter." (p. 61) This is the virginal aspect of Aphrodite (see
Appendix G: The Sacred Prostitute and the Erotic Feminine).

28. Also like Hermes, Hecate was a winged messenger (*angelos*), a guide
of souls represented by Hecataia built upon three cornered pillars at
crossroads (roadside Herms were 4-sided), had cakes and new smoked
offerings offered to her at every new moon, guarded gates and brought
wealth and good fortune to barns, and was associated with fruitfulness.
(Kerenyi 1976, p. 65)

29. The crassness Hermes shared with Hecate emphasizes his pre-Olympian
roots. *Olympian* Hermes relationship to indecency was perceived in a
different manner by Walter Otto: "But though the world of Hermes is
not dignified, and indeed in its characteristic manifestations produces a
definitely undignified and often enough dubious impression, yet—and
this is truly Olympian—it is remote from vulgarity and repulsiveness."
(Walter F. Otto, *The Homeric Gods*, Moses Hadas, trans. Thames and
Hudson: London, n.d., p. 123 quoted in Lopez-Pedreza 1977, p. 6.)

30. Marie-Louise von Franz quotes a Gnostic Ophite text describing Osiris
and his link to Hermes:

> [The] holy and sublime mysteries of Isis...have as their
> object nothing other than the phallus of Osiris...They
> say of the substance of the seed which is the source of
> all becoming that it is nothing in itself but produces and
> creates all becoming, since they say: "I become what I will,
> and I am what I am." Therefore is it that he is unmoved, yet
> moves everything. For he remains what he is, even though
> he creates all things and becomes no created thing...There
> is no temple where the hidden (that is, the phallus) does
> not stand naked before the entrance, erected from below
> upwards and wreathed with the fruits of all becoming...
> And the Greeks have taken over this mystical symbol from
> the Egyptians and kept it until today. We see, therefore,
> that the Herms were worshipped by them in this form.
> [Hans Leisegang, 1924, *Die Gnosis*, Kroners Taschenausgabe:
> Leipzig, p. 122 ff. (English translation from the German
> version of Hippolytos, *Elenchos (Refutatio Omnium Haer-
> esium)*, Vol. 2, In *Werke*, Vol. 3, edited by Paul Wendland,

1916, Leipzig. For English version, see: *The Refutation of All Heresies*, translated by J. H. Macmahon, 1911, Edinburgh, quoted in von Franz 1975, p. 25, 26]

31. The great primordial Goddess, mother of souls and mistress of ghosts, ruled over the island of Samothrace under the name Kabeiro. She was the primordial mother of the Kabeiroi, the gods who reigned on Samothrace. (Kerenyi 1976, p. 76) Hecate, another name for the great goddess, did not become associated with the underworld and the realm of souls until cleansed by the Kabeiroi and initiated into the underworld. (p. 43, 75) Hermes is prominently associated with the Kabeiroi mysteries due to the Herm as a phallic monument, the stone pointing "to the direct human experience of something divine." (p. 77, 78) The worship of Hermes as a phallic herm archetypically links him with the Kabeiroi. (see volume 2, Appendix H concerning Jung's phallic Self image) Taken together with their mutual associations of initiating the feminine into the underworld (cf. Hermes and Persephone), Kerenyi declares that it makes it "possible for the Herm to be considered the authentic symbol for the Samothracian mysteries. It is as God of the Kabeirian mysteries that Hermes is ithyphallic and guide of souls." (p. 75) In the Kabeiroi association, Hermes represents "the active and manifest original source and at the same time the prototype of a playfully and nimbly unfolding masculinity." (p. 81)

Norman O. Brown notes that close similarity between placements of the stone heaps associated with Hermes and Hermes as the phallus, discrediting the association of the phallus with fertility. For the Greeks, Romans and many others the phallus was thought to bring luck and ward off evil. The Romans used the term *fascinum* for phallus, meaning "enchantment," "witchcraft," emphasizing its magical abilities. (Brown 1969, p. 36, 37) It highlights the power, mystery, drive and fantasizing associated with the phallus but not reproduction. Ritual intercourse to make crops grow "presuppose[s] a body of biological theory that could have been acquired only through the development of agriculture beyond the rudimentary stages," making the application of the phallus "to vegetation magic...a late development." (p. 37, 38 note 5) Demeter and Dionysus were the Greek gods associated with vegetable fertility. (p. 35)

32. The ancient Greeks, Egyptians, Etruscans and Romans placed stone phalli on graves to symbolize "the *after-life of the spirit* and a guarantor of the dead man's resurrection":

> In ancient Egypt...the dead sun-god and king was honored in this way, as Osiris, and was represented by the phallic *djed* pillar. The erection of this pillar in the grave-chamber signified the resurrection of the dead man, who had become identical with the god Osiris. He was the green or

> black god of the underworld and also embodied the spirit
> of vegetation. (von Franz 1975, p. 24)

33. The begetter and the begotten are both present, even identical, in the masculine principle *per se.* (Kerenyi 1976, p. 78, 79) In the classical-Hellenistic Greek tradition Hermes was only represented as a child, the "divine child" and "son": the first born and the first begotten. In the preceding archaic period in association with Hermes Kabeirian nature, Hermes was depicted either as bearded or as a child (p. 81):

> In the mythologem of Aphrodite's birth, the phallus is also
> the child, just as Hermes is both the Kyllenic monument
> (the Herm) and the Kyllenic child. This identity receives
> its most tangible expression in the image of the paternal
> seed falling to earth in the form of a fruit...The ithyphallic
> pair (as the smallest number) in Samothrace represents the
> masculine in minimal unfolding. (p. 79)

> On Etruscan mirrors Hermes is called *"Hermes of Hades,"* expressing in the ancient Italian manner Hermes chthonic aspect and indicating the Hades-Hermes pair; Kabeirian father and son. (Kerenyi 1976, p. 80) For more associations of the phallus with the eternal source of life and therefore with the soul and eternal life (life after death) see Kerenyi 1976, p. 72.

34. The ram is a sacrificial animal and theiromorphic expression generally associated with the Kabeireans. (Kerenyi 1976, p. 85) "When Hermes begot Saos, the founding hero of Samothrace [the Kabeireans were the gods who ruled Samothrace], with Rhene (the 'sheep'), he certainly did so in the guise of a ram." (p. 85, 86) Hermes as a ram also begat the divine child of the mother mysteries. (p. 86)

35. [Hermes'] world originates before sunrise, and as source of his world he can only be the one

> who himself allows a source of illumination to originate in
> the outpouring of souls...The sun [is] reborn in every soul
> that is newly guided upward...As great Greek philosophers
> also knew, the source of light and the source of soul are one
> and the same. (Kerenyi 1976, p. 87)

36.

> Interpretation amounts...to an effort to tune the conscious
> attitude in such a way that a spark can fly out of the
> dream and over into consciousness, an "ah-ha" reaction is
> stimulated with a feeling of shock or illumination...Inter-
> pretation...is only correct when it seems "evident" to the
> dreamer, when it stimulates and also evokes an emotional
> alteration of the personality...One talks around [dream

images]...until such a reaction takes place. (von Franz 1975, p. 91, 92)

The classical description of dream interpretation is to reconnect consciousness with its source of energy, the archetype. This source of power is considered to be the primordial spirit that consciousness has differentiated itself away from and lost part of the primary energy contained in myth. The purpose of myth and dreams in this construct is to keep alive memory of our psychic prehistory down to the most primitive instincts. (Jung 1964, p. 98, 99) Assimilation of the myth's meanings broadens and modifies consciousness so as to heighten aliveness.

Archetypes have an emotional tone and the part of the brain associated with emotions is ramped up during dreaming. Interpreting a dream from an archetypal perspective connects consciousness with its emotional affective core and links the personal experience with the deep collective pattern (archetype) it is living out.

37. Hermes association with the Silenoi also link him with Dionysus since the Silenoi were devotees of Dionysus. A vase painting depicts the Silenoi in a drunken state playfully moving about wine goblets while their leader wears a Hermes' emblem of the traveler's mantel and herald's staff. (Kerenyi 1976, p. 89) These mythic associations depict the link between imbibing alcohol and the loosening of sexual inhibitions.

38. Alchemy is primarily concerned with a metaphoric and symbolic depiction of the phases and processes that occur in the space in Hermes' wand as the opposites are delineated and interact. These are the psychological processes that occur in the therapeutic relationship and/or personal development and transformation. An alchemical vessel is necessary to contain this process, the vessel being known psychologically as the therapeutic container or Winnicott's holding or facilitating environment. (D. W. Winnicott, 1960, The Theory of Parent-Infant Relationship in *The Maturational Process and the Facilitating Environment*, Hogart Press: London, 1965, p. 37-55)

39. Creation stories and other tribal myths are often told at the beginning of a Native American sweat lodge ceremony. The ceremony is meant to return one to the primordial beginning and a birth experience, re-minding participants of the mythic foundations of their psyches. This is one of many healing aspects of the sweat lodge ceremony.

40. Donald Winnicott pointed out the power in the space of objects close together (Winnicott 1966, p. 369) The paradox is that the more a sense of self one has, the more intimate one can be. (Scarf 1986, p. 75-76; discussed in volume 1, chapter 3 of *The Dairy Farmer's Guide*)

Poet Robert Bly has a good expression for the relationship of opposites: "Rejoicing in the opposites means pushing the opposites apart with our imaginations so as to create space, and then enjoying the fantastic music coming from each side." (Bly 1990, p. 175) This applies to multiculturalism, pluralism, democracy, honoring the value and attributes of each sex, and a sustainable relationship with the environment: one element is not increased at the expense and destruction of the other (Hexagram 41. Decrease).

The opposites of Hermes and Apollo depicted in Hermes' wand could be seen as Apollo representing cleanliness, purity, spirit, order, far-sightedness, reason, practice, left-brain hemisphere, archetypal masculine and light. Hermes would be more physical, indecent, unpredictable, chaotic, emergent, creative and more right-brain hemisphere; earthy, archetypal feminine, dark and unconscious. Linking the elements across the space in the wand would be communication, mediation, elucidation; the arts, music, mythology, etc., and a command or desire to be connected in a loving manner despite the fearfulness of the process.

Soul and psyche are born in the gap in Hermes' wand as are the arts and religion. The British psychoanalyst Donald W. Winnicott described the origins of religion and the cultural experience in the baby's experience of transitional phenomena with its first not-me object. (Appendix F) This in-between phenomena is Hermes' domain, with cultural phenomena being the initial experiencing of a *transitional object* that "doesn't stop." (Hogenson in Appendix G)

41. Raphael Lopez-Pedraza (1977) illustrates the importance of knowing the archetypal background of the different forms of sexuality in his examination of homosexuality. "Western culture has evidently lost contact with the archetypes which are behind eros among men," he observes, trying instead to understand homosexuality in terms of the father and mother. (p. 77) He criticizes both Jung and Freud for doing very little to explore Eros among men, hiding the conflict of eros/erotica between men. By knowing what god is alive in a male homosexual situation, "we can see through to the archetypal realm, the basic ground, from which the erotica of the whole personality stems." (p. 79) There is a misplaced tendency to see homosexuality "only in terms of illness to be cured or controlled" if the archetype is not recognized:

> What appears in the personal picture as "messes" (or illness, as psychotherapy tends to see it) are more likely expressions of the conflictive side of an archetype, or frictions brought about by the mixing of archetypes. Seen in this way, psychotherapy, by taking into account the archetype, could encourage psychic movement by following the dominant archetype in which erotica among men appears, and it would accept this dominant as the very vehicle for psychotherapeutic movement. (p. 78)

Lopez-Pedraza contrasts the tale of Apollo's servitude to Admentos to that of Hermes' service to Dryops, with the god Pan begotten from the Dryops' nymphal daughter:

> [Hermes' relationship] suggests the indirection of *falling in love with another man's fantasy*: the fantasy/erotica provided by the nymph. This could be the basis of what has been called a hermetic relationship between two men. This indirection, through a nymph belonging to the archetypal realm of Hermes, is in contrast to Apollo's direct and idealized conception of love among men. Apollo's archetypal realm allows us to perceive homosexuality in terms of initiation during adolescence. (Lopez-Pedraza 1977, p. 79)

42. Many powerful archetypal themes have been given imaginal form by a remarkable series of erotic drawings produced by Picasso in his late eighties. The drawings illustrate how sexual fantasies live on in old age, helping to keep the psyche alive and creative. Picasso's senility in old age is reflected in a series of erotic drawings of a Renaissance Cardinal as voyeur on the lovemaking of a painter and his model. Picasso's images can teach us about the psychology of the living madness of sexuality. His images reflect a profound historical and cultural conflict between Christianity and paganism, a conflict alive in Freud and Jung's works and indeed the collective psyche of Western civilization. (Lopez-Pedraza 1977, p. 70-73)

43. Raphael Lopez-Pedraza (1977) delineates three different aspects of sexual fantasy associated with Hermes chasing a nymph; chasing, catching and rape. "In this chasing of the nymph, we shall never know—in psychotherapy, in life, or in...Don Juan [or]...Marquis de Sade...if the archetype is going to fall into a soul-making process, or so-called psychopathy, and even into criminality...Implicit in the fantasy of rape is that borderline in which either soul-making or delinquency can occur." (p. 106) This is Hermes "leading the way or leading astray."

Lopez-Pedraza's exploration of Hermes chasing the nymphs has ecological implications. The classic story of the rape of a virgin is traced to Cicero's tale of a love-affair between Mercury (Hermes) and Diana (Artemis), the principle of religious virginity, resulting in the birth of Cupid (Eros). A dark dimension of this archetype is Don Juan trying to meet his erotic, archetypal challenge, moving not into Eros but dishonor and murder. Of the three classical Greek virgins, Hestia (virgin of home and hearth—privacy?), Pallas Athene (virginal side of *polis*—public life) and Artemis (virginal nature),

> Artemis...most arouses the fantasy of being chased, a chasing that can move into rape. She is the wildest, her realm is nature and hunting, the untouched virginity of nature. She is lunatic (moon-like), and has her own touch of religious fanaticism. The iconography portrays her as a boyish maiden, and this hermaphroditism is an added ingredient

of her sexual attraction within the tension created by her
being a virgin Goddess. (Lopez-Pedraza 1977, p. 109)

Euripides focuses on the virginal Artemis "as a blood-thirsty,
demander of victims of sacrifice," perhaps an aspect of the destructive
side of Great Mother nature. To Lopez-Pedraza it seems "the virginity which most arouses the fantasy of being chased is precisely the
complementary partner to killing and destruction." (Lopez-Pedraza
1977, p. 115 note 35) Virginity is also a basic motif in Sade's work, a
man intoxicated with 18th century natural philosophy. (p. 115 note
34) Lopez-Pedraza notes how our culture has fluctuated between the
chastity of Artemis and the sexual desire of Aphrodite. (p. 74 note 7)

APPENDIX A

Dynamic Systems Theory

James Gleick's publication of *Chaos* in 1987 described a revolutionary paradigm of interest to Jungians and ecopsychologists. Chaos theory focuses on "the irregular side of nature. The discontinuous and erratic side" that has long been a puzzle to science. (Gleick 1987, p. 3) It is "a science of process rather than state, of becoming rather than being." (p. 5) Psychologist Frederick Abraham sees dynamic systems theory (DST), closely related to chaos and complexity theories, as an archetype, a meta-language for "describ[ing] the complex interactions of the multiple aspects of the psyche and their evolution over time, to combine the *analytic* and the *holistic*." (Abraham 1995, p. 65) Complexity theory is a mathematical description of how new possibilities are created in complex systems that, at the psychological level, offer conscious choices that lead to individuation. These theories offer a mathematical description of basic processes in nature that humans experience emotionally and symbolically and as a transformative narrative.

As a science about the global nature of systems, complexity theory cuts across scientific disciplines and reverses the tendency towards specialization. "It makes strong claims about the universal behavior of complexity," Gleick writes, and reveals patterns that appear "on different scales at the same time." (Gleick 1987, p. 5)

These statements echo Jung's interest in the archetypal (basic patterns), the complexity of the psyche if one goes beyond a focus on the ego, the wholeness of the Self, the irrational, the feminine as process, the earthly and the natural. Chaos theory destroys several basic tenets of reductionist science, "dismiss[ing] the idea that with the right equations and enough data, the behavior of the most complex system could be predicted." (Schmidt 1995, p. 11) Systems like weather, population dynamics, epidemics, fluid dynamics and economics don't fit linear systems—those that can be graphed with a straight line, are solvable, and can be dissected into pieces and put together again: "Nonlinear systems generally cannot be solved and cannot be added together." (Gleick 1987, p. 23, 24)

Sensitive dependence on initial conditions is a fundamental element of chaos theory. Very small initial differences get magnified exponentially over time into global uncertainties, making it impossible to make long term weather forecasts for example. The theory also describes how many unpredictable and unstable dynamics usually operate within a system bounded and governed by laws. They never settle down to a fixed value or repeatable pattern, but neither do they move off into infinity. Such systems are called strange attractors. Attractors are the hidden global forces that shape the overall behavior of a system. "An attractor is the pattern of behavior to which a system 'settles down' or is attracted" (Crutchfield et al 1986 referenced in Schmidt p. 12) much like an archetype is described.

Chaos theory has a mathematical base whose formulas produce graphical results. Plotting the results of running different numbers through equations associated with strange attractors reveals geometric patterns (fractals) that have the same structure going from a small scale to a large scale (scalar invariance), with branching twigs having the same pattern as the major branches of a tree for example.

Even some of the simplest interactive functions produce characteristics of dynamic systems. Increasing the energy or rate of activity in a logistic system results at first in two equilibrium values. Further increase in the rate causes each new value to split, beginning a process of bifurcation "into progressively more potential values." (Schmidt 1995, p. 13) Complex system functions are revealed when the rate increases enough that an infinite number of values become possible. "In effect, driving the system harder increased its non-linearity and its potential for producing different and unpredictable outcomes." (p. 13, 14) The fact that the doubling has a constant rate reveals the universal nature in chaotic processes. (p. 14)

When an open system receives energy and becomes unstable, *it self-organizes by producing new patterns* and complexity. (Kaufman 1993; Levin 1992, Barton 1994 in Schmidt 1995, p. 14). "Only chaotic states will respond to stimuli by producing new information in the system." (King 1991 in Schmidt 1995, p. 14) Complexity theory thereby provides the mathematical base for the generative, creative processes in nature and in the psyche. With dimensionality being an indicator of the degree of complexity, EEG recordings of oscillations in brain activity revealed higher dimensionality in subjects involved in the more complex information processing associated with imagining an object than in those who actually perceived an object. (Lutzenberger et al 1992 in Schmidt 1995, p. 16)

Freeman's EEG recordings from electrodes implanted in the olfactory bulbs of rabbits "found that every neuron in the bulb was involved in every stimulus response and that it was the spatial patterns of amplitudes over the entire bulb that distinguished one stimulus response from another." (Schmidt 1995, p. 17) The brain generates background chaotic activity in the olfactory bulb and the olfactory cortex "by the mutual excitation of the two regions." (p. 18) Rapid, global changes of state moving from low to high dimension attractors appear "in response to weak olfactory stimuli, i.e., bifurcations" producing the classic pictures of strange attractors from computer simulations of EEG patterns. (p. 17, 18) A novel odor will avoid existing attractors as olfactory bulb activity "enters a high dimension chaotic realm until a new attractor is established." (p. 18) Freeman's (1990) detailed neuroactivity maps showed that perception is established by the "deterministic chaotic activity of all neurons instead of discrete patterns of activity of specific neuronal circuits." This led him to propose that "the ability to create activity patterns may underlie the brain's ability to generate insight." (Freeman 1991 quoted in Schmidt 1995, p. 28)

Freeman (1991) argued that "chaos underlies the ability of the brain to respond flexibly to the outside world and to generate novel activity patterns, including those that are experienced as fresh ideas," and that chaos "may be the chief property that makes the brain different from an artificial intelligence machine." (quoted in Schmidt 1995, p. 18)

The uniqueness of each human mind may occur because in the developing brain there is a fluid and dynamic growth of the neuronal axons that connect one neuron to another. This presents the opportunity for every brain, as for every creature and every snowflake, to be unique:

> Environment, translated through the senses, strengthens some synapses and can affect actual brain architecture (Hubel 1977; Kandel 1989). The response of axonal growth cones to the microenvironment of the developing brain may have a parallel in the creation of snowflakes... with their infinite variety contained within the bounds of hexagonal symmetry. As described by Gleick (1987), each snowflake displays sensitive dependence on initial condi-

tions. Each one experiences a different interplay between heat diffusion and surface tension, and, as a result, becomes a unique creation. (Schmidt 1995, p. 20, 21)

Stable human traits are not hard wired for that would "imply linear, point-to-point, reductionistic engineering." (Schmidt 1995, p. 21) In Freeman's dynamic systems "in which every neuron participates in forming recognition patterns...stable attractors are formed by perceptions and strengthened by repeated combination with other relevant stimuli." (p. 22) This creates stable personality traits, yet the brain can create new response possibilities (Mpitsos 1989) through the process of bifurcation (Milton 1989; Zak 1991):

> Lewin (1992)...suggest[s] a tendency for complex situations to be poised at "the edge of chaos" (Nicholas 1986), the mathematically defined region at which state transitions occur among bifurcated outcomes. In this region, a system is optimally adaptable and creative. A chaotic system is always capable of escaping from stable states under unfamiliar stimulus and creating new states. (King 1991 referenced in Schmidt 1995, p. 22)

Perhaps perception can occur quickly and accurately, "even when the stimuli are complex and the context in which they arise varies," because "the brain uses processes of inference, interpretation, and adaptive logic unlike the deterministic algorithms of the computer." (Uttal 1990 referenced in Schmidt 1995, p. 22, 23)

Dynamic systems may explain many aspects of memory:

> An attractor system contains far less data than required to describe all dynamical states...(King, 1991)...Alkon (1989) suggests awareness of a small portion of a memory pattern will trigger awareness of the whole. Any stimulus that triggers the spatial pattern of neural activity producing a particular attractor will generate the full richness of the remembered experience. (Schmidt 1995, p. 23, 24)

Some memories are more strongly recalled because "in Freeman's model, simultaneous pairing of an amygdala-generated emotional response with a stimulus would strengthen the synapses involved in the attractor which stored a particular memory." (p. 24)

The operative neuronal circuitry in even the simplest neurosystems is impossible to define. (Sejnowski 1988; Uttal 1990; Mpitsos 1989 referenced in Schmidt 1995, p. 24) By contrast, psychiatrist Gregory

Schmidt notes, "The concept of strange attractors involving the activity of all neurons in an interconnected network and simultaneous activity of multiple attractors in different brain areas activating each other provides an intuitive model for the complex, instantaneous integration of function which occurs in the brain." (p. 24)

Learning may occur "because chaos provides the brain with 'a deterministic "I don't know" within which new activity patterns can be generated.'" (Skarda and Freeman (1987) quoted in Schmidt 1995, p. 25) [the gap in Hermes' wand as described in volume 3 of *The Dairy Farmer's Guide*]:

> All learning can be postulated to occur in this way with creation and strengthening of new attractors...Chaotic processes amplify small fluctuations (Crutchfield 1986) and create new activity patterns (Freeman 1991). The development of new attractors, whether from new information or activation of repressed information, would be expected to alter existing attractors in a way which could generate new understanding. (p. 25)

The dreaming brain operates on the principles of complexity theory and repressed information is more easily accessed and portrayed in imaginal and story form in dreams. These factors are largely responsible for the creativity associated with dreams and their usefulness in the psychoanalytic process.

Frederick Abraham recognized the significance of DST in describing basic psychological and Jungian concepts:

> Mathematical dynamics provides a meta-modeling format to describe how multiple forces interact. As with the psyche, some of these forces pull or push for *convergence* to attractive regions, while other forces may push or pull for *divergence* away from these regions. Pictorially, the potential paths (*trajectories*) created by these forces comprise a *portrait*. *Control* factors may change these forces a bit, which in turn would reorganize the pattern of the portrait and its *attractors* dramatically in transformations called *bifurcations*. (Abraham 1995, p. 66)

Self-organization, a quality of dynamic systems, arises because bifurcations allow a system to influence its own controls. This is illustrated by *enantiodromia*—a reversal to the opposite as in Jekyll and Hyde—and *individuation* through choices made at bifurcation points. Dynamic systems also reveal an interconnectedness among component elements. (Abraham 1995, p. 66)

Using italicized words for the mathematical vocabulary, Abraham illustrates the application of DST to everyday psychological experiences:

> During *catastrophic* storms of *transformation* (*bifurcation*) in adolescence, most young people go down many *pathways* (*bifurcation*) that lead to what has been called "the generation gap." Parents are often shocked at the *chaotic* (a.k.a. *strange*) *attractors* that seem to draw their children into unexpected developmental *trajectories* in their attitudes, romantic entanglements, and careers. Parents hope the sensible *initial conditions* they set up will be *deterministic* in facilitating a trajectory to a nice destiny (*stable*, mature *attractor*) following their own cognitive models. Yet life is filled with so many *complexities* that many parents just throw up their hands when they finally realize they really cannot predict or control just what their kids will do. "That's life!" they say. (Abraham 1995, p. 66)

Many of the concepts of depth psychology can be framed in DST terms. Alchemy, fairytales, myths and mandalas are studied "with the hope of learning to recognize ground plans (*dynamical systems, phase portraits, dynamical schemes*, and *response diagrams*) that will illuminate the archetypal patterns (*attractors*) underlying the individuation (*bifurcation*) of their patients." (Abraham 1995, p. 67)

Abraham examines determinism and free will in the context of the courage and self-realization necessary to deal with chaos and bifurcations in order to achieve individuation. He notes, "The word *chaos* originally came from the Greek word meaning 'gap,' the *creative* void that gave rise to Gaia (Mother Earth) and Ouranos (Father Heaven), or the gap between them from which sprang all else." (emphasis added) This is the gap in Hermes' wand (volume 3). The word later got associated with the disorder connected with darkness and the obscene elements of the Greek underworld, associations similar to those in Icelandic and Babylonian creation myths. (Abraham 1995, p. 76)

Chaos in mathematical dynamics (DST) terms refers specifically to the complex or chaotic trajectories of chaotic attractors, with great complexity being associated with high dimensional systems and low complexity being associated with low dimensional systems. But, "like all attractors, they are stable and governed by order, no matter how complex." (Abraham 1995, p. 67, 68)

The old meaning of the *gap*, in DST terms, is the bifurcation point "at which the attractor that is in the process of transforming does not exist." There is no chaotic attractor at this point: "There is neither egg nor chick," writes Abraham. "[It is] the fearful, creative abyss, the gap, such as *ayin* in Jewish mysticism." (Abraham 1995, p. 68) "As the bifurcation occurs, some portion of the attractor appears, disappears, or is replaced by a new, different attractor." The bifurcation point is very unstable:

> A slight alteration of a force can tip the system back to the previous attractor or lack thereof, or on to a new attractor or loss thereof, in the portrait. However, it takes *energy* to change the control in order for this to happen, to create this small change in the balance of conflicting, cooperating, and interacting forces. (p. 68)

Bifurcation points offer the possibility of conscious choices of direction that lead to individuation:

> From the mathematical dynamic point of view, one type of choice is seen as resetting of the "initial position" into a different *basin of attraction* in order to favor the trajectory tending to a particular goal. A more active type of choice is to navigate along the controls of the whole potential *dynamical scheme,* learned or imagined, to create new possible attractors of self. For the psyche, this is individuation, a process of bifurcation leading to a greater complexity and maturity. The psyche, as a dynamical scheme, views potential worlds and chooses among them. (Abraham 1995, p. 68)

Polarities in the human psyche and the importance of maintaining the tension of the opposites are important Jungian concepts. Jung saw a great divide (therefore a lot of energy) in the American psyche between its European civilized consciousness and the aboriginal land and cultures upon which it was superimposed. The pathological aspect is addressed by James Gustafson in the University of Wisconsin-Madison Psychiatry Department. He speaks of the epidemic illness of the modern mind as being the consequence of its linear orientation being ill prepared for the huge set of non-linear bifurcations in life. Gustafson conceives of the aboriginal mind as being "an integrative mind, transitional between all the great opposites of nature, like hot and cold, dry and rainy, light and dark, fresh and rotting, and so forth. Put such a mind into the modern world of linear programs (schooling, assembly lines...etc.) and you have an isolated linear will disconnected from all its opposites and alarmed by all of them." There is a significant "bifurcation of the head (the leading part of the personality, or mask) from the body (in shadow), the abstract from details, scientific will from romantic will...etcetera." (abstract from an address given on April 24, 2001 to the Chaos and Complex Systems Seminar at UW-Madison)

Being able to stay in the liminal space that is the bifurcation point challenges the psyche/system to present many new possibilities offered in the form of dreams, visions, insights, creative ideas, new behaviors, a new worldview, a different way of framing things, etc. Numinous images of Self often occur at these moments of great duress when the individual is in the "don't know" space and desperately in need of an integrative perspective. The gap is often frightening and unnerving: one can feel totally ungrounded and lose a sense of identity. There is a temptation to revert back to old defensive structures (old stable attractors) to fend off fears, doubts, uncertainties and to avoid re-experiencing old wounds. The analytic process can offer a container, the alchemical vessel of transformation, by the analyst becoming an ally accompanying the analysand into the gap as the analysand searches for their inner truth and a path with heart and meaning in their lives (new attractors and new trajectories). (see volume 1, Appendix B: The Alchemy of Psychoanalysis) The Chinese ideogram for "crisis" is composed of the elements "danger" and "opportunity."

> Danger itself
> Fosters the rescuing power.
> Holderine (quoted in Abraham 1995, p. 65)

James Hillman's re-visioning of our relationship with the objects in the outer world offers an avenue for increasing complexity. He promotes the neo-Platonic concept of Aphrodite as the Soul of the World (*anima mundi*) and suggests that we approach the world with a feeling-imaginative heart instead of with the head (Appendix K). The resulting fascination with the beauty and complexity of the object, human or natural, draws one into a more intimate relationship with it. Fascination, awe, enchantment and the numinous, in complexity theory concepts, are high energy experiences which generate high dimensionality/complexity. Immersing oneself in the complexity of nature has healing (moving-toward-wholeness) potential—one can see one's life and problems reflected in a transpersonal framework, can see the storm in the valley from the mountaintop. The ultimate intimacy and intensity, however, comes from engaging the Self in a personal, emotional manner as Jung did in writing "Answer to Job."

APPENDIX B

Bootstrapping the Archetypes

George Hogenson, a leader in applying dynamic systems theories of development, situated robotics, and non-Darwinian theories of evolution to depth psychology, argues that "archetypes [and the collective unconscious] do not exist in some particular place, be it the genome or some transcendent realm of Platonic ideals. Rather, the archetypes are the emergent properties of the dynamic developmental system of brain, environment and narrative." (Hogenson 2001, p. 607)

This concept is in stark contrast to the commonly held theory of archetypes presented by a writer like Anthony Stevens. This model is based on the computer metaphor used in cognitive science and artificial intelligence stating that evolution has produced a mind that "contains representations of [typically encountered] states of affairs in the world" and an accurate view of the world is derived by "applying various computational processes to these representations." (Hogenson 2000, p. 13) "The mind is taken to consist of a large number of highly defined, innate modular components [the innate a priori archetypes] that guide the major forms of behavior." This approach has been unable to model such elementary activities as walking. A basic problem is "that sensory inputs were first passed through an elaborate cognitive module which then had to initiate motor inputs." (Hogenson 2003, p. 108)

I. Situated Robotics and Archetypes

Situated robots are part of a paradigm shift for understanding the mind and behavior. A robot designed to inhabit a particular environmental situation will "display unusually independent and unprogrammed forms of behavior" that emerge out of programming the robot to perceive and respond in simple ways. (Hogenson 2000, p. 3) "Robots with hardly any computational abilities...were able to traverse relatively complex environments." (Hogenson 2003, p. 108)

For example, Rolf Pfeifer constructed a square enclosure six feet on a side containing about thirty 2" x 2" polystyrene cubes. Three small bumper cars were programmed to turn to the right whenever they sensed a barrier, i.e., bumped into something. Herding behavior "emerged"—within 20 minutes the robots had arranged the cubes into two or three clusters. "The behavior that one observes emerges in the setting in which these robots exist, and for which they were designed." (Hogenson 2000, p. 4) Herding behavior was not the result of an elaborate computer program based on a cognitive science model that works in modular packets to calculate if what was perceived matched an inner image (one module/sub-program) and then moved the robot in complex ways (second module) in conjunction with the activities of other robots (third module) to group/herd the cubes into a programmed image of the final result (fourth module).

In dynamic systems theory the mind arises from the self-organizing evolution of simple behaviors linked to perceptions in given situations: there is no template called archetypes in the brain, the genes, or a Platonic realm of ideals. (Hogenson 2004b, p. 71) Archetypes emerge, are bootstrapped, with information being fed to the brain by simple human perceptions and responses to the world which configure the brain to carry out processes "on an increasingly organized and systematized basis." This is demonstrated by robots in a situated environment. Without a pre-existing template in the system, "a robot that starts out using simple sensors to navigate a room can pass off its parameters to a neural network, which will, over time, take over the navigational process, displaying such features as anticipation of known objects and other developmentally sophisticated behavior." (Pfeiffer and Scheier 1999 summarized in Hogenson 2004b, p. 73) The neural network is connectionist modeling as demonstrated by using massively parallel computation. Machines can proceed with little or no programming to be taught, for example, "basic grammatical patterns in relatively short time." This disproves Noam Chomsky's theory "that grammar had to be innate because it was too complex to be learned." (p. 69)

Language acquisition offers a prime example of the difference between the classical view of archetypes and archetypes as emergent phenomena. Linguist Noam Chomsky maintains that young children in all cultures easily and rapidly learn their native language because of "innate structures in the brain that defined a set of syntactical rules—a

deep grammar—common to all languages." (Chomsky 1965 summarized in Hogenson 2001, p. 603) But languages change too rapidly and
there are too many basic differences in the fundamental elements of
language to be accounted for by the neurological capacity of the brain.
In addition, the human environment in which languages developed
was not stable for a long enough period for genomes to be selected for
the entire human race. (Deacon 1997, p. 333 ff referenced in Hogenson 2001, p. 604) The traditional concept of archetypes seems more
improbable if there are considered to be as many archetypes as there are
typical life situations. (CW 9, I, ¶ 99)

The situated environment of a human infant is of total dependence
on intimate interactions with an adult caretaker, a limited set of percep-
tions/responses, and a large, programmable brain. DST examines the
interaction between human structures and the adult-infant interaction.
For example, there are several significant factors affecting the human
capacity to learn a language. Humans have many more neural connections than other primates between the brain and the throat and tongue
(Hogenson 2003, p. 109) and an auditory system that can transmit much
more complex patterns. (p. 113) This enables more subtle distinctions
in the ability to hear and produce sounds. Poor short-term memory
in children facilitates language learning (Hogenson 2001, p. 604) and
"languages have structures like the noun/verb distinction because the
brain [has] structures that make this form of categorization more effi-
cient for learning and using the language" (p. 605):

> Children do not simply begin to spontaneously generate
> grammatical language [due to some specifiable neurologi-
> cal structures in the brain] but rather go through a very
> particular learning process heavily dependent on stereo-
> typical interactions with their parents or care-givers...
> These stereotypical *learning processes*...are more likely to
> be stable over time and space, and...the co-evolution of
> language and brain relies on them to establish our linguis-
> tic abilities. (p. 604, 605)

Hogenson notes, "Some higher-level functions may end up being
weakly modular...by virtue of the self-organization of the brain's
neural network" that emerges during a developmental process, but
these modules do not exist a priori in the neocortex as some inherited
interconnection of neurons, innate archetypal pattern, etc. (Hogenson
2003, p. 109, 110) Neuroscientist and anthropologist Terrence Deacon
acknowledges that grammatical universals do exist but as a result of

convergent features of language evolution: "[Language universals] emerged spontaneously and independently in each evolving language, in response to universal biases in the selection processes affecting language transmission." (Deacon 1997 p. 115f quoted in Hogenson 2001, p. 605)

II. Myth and the Symbolic Universe Humans Inhabit

The human mind is situated in the basic structure and functioning of the human brain located in a human body with fundamental differences between the sexes. (see Appendix C: The Human as an Embodied Robot) Physiologically the body needs energy derived from food and has to live within the parameters of all living organisms such as a limited temperature range (neither too hot nor too cold). It has to be protected from dangerous natural elements like avalanches and poisonous snakes and manmade dangers like vehicle crashes. The natural environment was the primary selective force up to about four million years ago when "the early hominids such as Australopithecus—the famous Lucy—had diverged from the other apes." (Hogenson 2000, p. 19) This is on the border between level F—the primeval ancestors, in Jung's diagram of the collective unconscious (volume 1, p. 35) and the level beneath it, "the animal ancestors in general." Jung called this still living part of our psychic inheritance "the primate within."

The evolutionary environment rapidly shifted to an environment of symbols and artifacts with the emergence of the true line of Homo, in Homo erectus, that displayed sophisticated tool making. (Mithen 1996 referenced in Hogenson 2000, p. 19, 20) Humans became a symbol using creature living in "a natural environment of meaning" that subsequently shaped the organic development of the species. (Hendriks-Jansen 1996, p. xi quoted in Hogenson 2000, p. 20) Further evolution of Jung's "two million-year-old man within" came to be "largely governed by the demands imposed by the symbolic, cultural, environment." (p. 20)

Patterns of human interaction have selectively evolved that best facilitate human development. This is a co-evolution model in which artifacts generated by humans, including cultural forms like myths, are at least as important as the evolution of human mental abilities. This was experimentally demonstrated by modeling "processes by which a culture could acquire the capacity to predict an event that no individual could learn to predict in a single lifetime" such as correlating moon phases with tidal patterns so an inland foraging tribe would move to the sea at the appropriate time to harvest mollusks. Organic evolution over a long period of time produced a population better at predicting, but "much greater success was achieved...when all of the agents in a given generation produced an artifact that could be used to predict the tides. The next generation was then set to select the best artifacts and attempt to improve on them." (Hogenson 2004b, p. 74)

DST can also be applied to a higher scale of human development—what Jung called the mythopoetic dimension of the human psyche, the dimension through which the gods have traditionally spoken. Philosophical and anthropological research has led to an increasing acceptance of an evolution in cultural artifacts including myths. (Hogenson 2004b, p. 74, 75) Deacon proposes that language in general (and Hogenson adds, "specifically those stories that last, i.e., myths"), "has evolved to work within the brain and developmental setting of the human infant." (Deacon 1997 referenced in Hogenson 2004b, p. 75) Myths and symbols are part of a dynamic mix of interactions "of agents, environment and other artifacts" that facilitate the emergence of an adult with intentionality, meaningful action and relationships. (p. 75) Because the surface manifestation of symbols and myths is so vastly different, like languages, there is probably not an evolutionary base for a common brain structure.

Another important concept in DST is the dynamic linking of time frames. Time frames, seen as a process in time, are contained or nested within larger processes extending over longer periods of time. Hogenson's example: "The real time process of infant-caregiver interaction... [influences and] interacts with developmental time" in turn nested within cultural and historical time "and finally to symbolic and mythic time." (Hogenson 2004b, p. 72, 73)

III. The Analytic Process and the Self from a DST Perspective

DST offers an alternative theory about the stages of life, the analytic process and the emergence of the Self. "Beginning at least with Freud, human psychological development has been largely conceived of in linear terms," notes Hogenson. "A series of stages were postulated" whose progression could be inhibited in part by such events as traumas, resulting in various forms of psychopathology. However, there is no intrinsic structure in the stages nor usually is there any real goal the individual is working toward. The individual is forming a competency (basin of attraction) around a particular task (attractor). In DST terms the individual is said to be moving through a phase space in various stages analogous

> to the movement of a pendulum through the range of its swing. The range of the swing, in turn, defines a basin of attraction for the movement of the pendulum. Any perturbation of the pendulum alters the phase space, and in turn defines a new attractor. What is seen as a stage in classical theory is seen as a relatively stable attractor in dynamic theory. (Hogenson 2004b, p. 75, 76).

Trauma is recast within this framework as a new attractor/influence.

In the psychoanalytic process described in DST terms,

> "[the] apparent developmental regression [results from an] increased systemic variability at phase transitions in the dissolution of one stable attractor as the system moves toward another...The system appears to become less complex, more disorganized. The system is also more sensitive to disruption or trajectory changes at these points. However, following this brief, variable period, the system will reorganize, and the 'missing' behaviors may spontaneously reemerge. Usually they will be more stable, reliable, and more complex than before the reorganization." (Tucker and Hirsh-Pasek 1993, p. 366 quoted in Hogenson 2004b, p. 76)

In analysis most of the interpretations and analytical decisions occur at the phase transitions "seen as regressions, splitting or projective identifications." There is also

> an entrainment of posture, tone and other interactive aspects of their treatment experiences. This is the essence of the interactive approach, but it is also the case that movement is not back and forth along a linear path, but, as it were, throughout a multi-dimensional space of continually changing phase transitions through state space. (Hogenson 2004b, p. 76)

From this perspective Hogenson notes that the self cannot be seen "as some stable state of affairs that can, even in theory, be attained." (p. 76) "With the Self as a way of conceptualizing wholeness (CW 9, II)" and

Jung's contextual sense of image "taking into account the entire setting of the organism's life and behavior (Hogenson 2001)," the self in DST terms becomes "a super-ordinate principle that overarches the system of the psyche, and even the system psyche world." (Hogenson 2004b, p. 77) The self expressed as a god image "defines the entire dynamic context, implicit and explicit, within which the individual is acting. The god image is, in other words, the phase space of the individual's life organization." (p. 77) "The self...[is] the sum of the available attractor states within phase space through which a process of self-organizing emergence can take place at any given point in time." (p. 76)

In the end one's sense of self is the virtual reality of a symbolic self. "Consciousness of self," Deacon remarks, "implicitly includes consciousness of other selves, and other consciousnesses can only be represented through the virtual reference created by symbols." It is a symbolic self "that is the source of one's experience of intentionality," judgments and fear of death. "It is a final irony that it is the virtual, not actual, reference that symbols provide, which gives rise to this experience of self. This most undeniably real experience is a virtual reality." (Deacon 1997, p. 452 quoted in Hogenson 2004b, p. 77) Hogenson adds, "by virtue of the symbolic, the human world becomes an infinite world, where the sense of oneself is a constantly emerging sense of potential states and phase transitions that have yet to be traversed." Hogenson astutely emphasizes that psychodynamic depth psychologies may be the most dynamic systems because Jung argued that in the deep transformations in the psychoanalytic process the analyst must be prepared to be transformed as well. (Hogenson 2004b, p. 78)

DST from a Jungian perspective is that "the presence of simple patterns of perception and action, occurring in species typical environments and enlisting species typical forms of interpretation, will be seen to give rise to the immense beauty and complexity of the great myths of our species." (Hogenson 2000, p. 23) Alchemy and the myths symbolically depict the attractors and alchemy in particular symbolizes with archetypal imagery the processes accompanying the transition states.

IV. Reframing Jung's System from Word Association to Synchronicity

Hogenson has reconceptualized Jung's system of psychology using DST, the self-organizing processes in complex systems, and other concepts. He points out that the relationship of many natural phenomena come into existence

> by virtue of the dynamics of the system in which they were imbedded and which they helped form. Self-organized systems, while made up of many elements—commonly referred to as a complex system—nevertheless display high levels of organization regardless of the scale at which they are examined. (Hogenson 2005, p. 273)

The particular pattern that emerges in any self-organizing system is unique and largely unpredictable, yet "the structure of the systems in a given domain will display a great deal of what is called self-similarity [fractals]." Every individual for example has their own set of associations or semantic networks to particular words, yet "semantic networks have features of self-similarity regardless of whether one is reading a child's book or the *Critique of Pure Reason*. In the case of language, if self-similarity in the system did not exist, communication would become impossible." (Hogenson 2004a, p. 4)

An important aspect of intensely self-organizing systems is that if they go on long enough they "reach a point known as self-organizing criticality...[or singularity]...[where] even a small deviation from the organizational pattern can cause the entire system to reorganize itself in an abrupt and unpredictable, even catastrophic, manner." This is illustrated by slowly pouring grains of sand onto a table. A cone-shaped pile soon emerges that appears to be well ordered. However, at some unpredictable point, "the introduction of one more grain set[s] off a 'catastrophic' reorganization of the pile...a dramatic 'avalanche' on the side of the pile." (Hogenson 2005, p. 274) This is the point of self-organizing criticality that emerges from the self-organizing properties of the system. (p. 275) Another example can be seen in the stock market. The major traders progressively self-organize by entraining to each other's behavior until they behave in a tightly defined pattern that causes bidding to escalate exponentially. A sudden change in the trading pattern by only one investor can bring about a collapse of the whole system. (p. 274). The stock market, like the sand pile, has undergone what is called a phase transition, an abrupt change of state as when water freezes.

Hogenson proposes that Jung's system of word association, complex, archetype, the emergence of the self and a synchronistic event is a continuum of emergent self-similar (fractal) structures within a symbolic system. The increasingly complex structures display a power law distribution (mathematical qualities of the geometric structures of fractals) and arise "through a series of self-organized critical moments that result in phase transitions within the symbolic system as a whole." (Hogenson 2005, p. 278) As "symbolic density" increases, the self-similar structure transits to the next phase as a "symbolic avalanche" occurs on the steep sides of the symbolic sand pile and radically reorganizes one's world. (Hogenson 2004a, p. 16) Within the symbolic system that is human language, Kant's *Critique of Pure Reason* is a child's story that didn't stop. (p. 12)

Single words or other symbols can have an enormous range of associative connections unique to the individual as Jung demonstrated with his word association experiment. A complex is described as a structural pattern in a feeling toned group of associations of memories, words, behaviors, etc. and are felt to be a personal experience. In Hogenson's schema, archetypes don't form the complexes as Jungians usually describe it, but archetypes emerge out of complexes as experiences move from a sense of the personal to the objective, collective, and impersonal realm of the archetypes. "The complex and the archetype are fundamentally structured [fractal nature] like the symbol, only the archetype exhibits itself at the point where symbolic density transcends the carrying capacity of the complex and moves into a more collective realm." This occurs at "an iterative moment in the self-organization of the symbolic world." (Hogenson 2005, p. 279)

Jung's system is based primarily on the nature and function of the symbol, making it necessary to understand symbols if one is to understand the psyche. This begins with Jung's emphasis on the autonomy and creativity of fantasy, the two being linked. The autonomy of fantasy stems from it being "the mother of all possibilities," meaning it is endlessly creative with the creative activity of fantasy being expressed in symbol formation. (CW 6, ¶ 78) "Fantasy as imaginative activity is...simply the direct expression of psychic life," Jung wrote, "[and it] appear[s] in consciousness...in the form of images and contents" that have an energy or force behind them, an idée force. (¶ 722) Fantasy pervades all psychic acts, drawing together and mediating all aspects of the psyche (¶ 722) and mediates our experience of the inner and outer world. (¶ 78) Jung said fantasy can appear in primordial form or as "the ultimate and boldest product of all our faculties combined." (¶ 78) Psychic energy is increased in dealing with "irreconcilable claims of subject and object, introversion and extraversion," and it is fantasy that bridges these and all opposites (¶ 78) through its symbol-creating abilities, with symbols being described as a combination of the known with the unknown.

Terrence Deacon uses mathematical analogues as did Jung (Jung 1961, p. 310, 311) to describe the absolute autonomy and formal restraints on form and structure that seems to characterize symbols. Jung said numbers were the purest form of the archetypes and best illustrated the "just so" nature of archetypes. (von Franz 1975, p. 126) Deacon compared symbols to prime numbers in mathematics that exist independently of the brain or anything else. (Deacon 2003, p. 98 in Hogenson 2005, p. 279, 280) Prime numbers are numbers that cannot be divided, like 17, whereas 16 is not a prime number because it is dividable (by 2, 4 and 8). Hogenson describes Deacon's argument: "Even if nobody ever calculated a prime number the system of primes would nevertheless be said to exist, and the process of mathematical investigation becomes one of discovery rather that construction, at least in part." Hogenson suggests it is "possible to conceive of the world of the symbolic as a world that the psyche inhabits and realizes, or perhaps falls into, rather than as a world that the human mind creates." (Hogenson 2005, p. 280)

Jung located synchronicity within the context of the symbolic, usually associated with archetypes. A profound affective sense of meaning that can change a person's life is felt when inner and outer worlds are juxtaposed in a synchronistic event. Hogenson believes synchronicity must be taken seriously and seen "as an organic and in fact perfectly reasonable element of a system of psychology that takes the symbolic and the creation of meaning as its central and guiding principles" (Hogenson 2005, p. 273):

> If we see the symbolic as more than simply a system of representations but rather a relatively autonomous self-organizing domain in its own right...[then] the complex, the archetype, the synchronicity and the Self all 'exist' as moments in a scale invariant distribution governed by a power law...Synchronistic phenomena are extremely rare...but they are not improbable in the sense one would

assume to be the case with more conventional probability theory. They are rather the result of small symbolic developments that do not stop. (p. 281)

Hogenson sees synchronicities as being exceedingly rare "[but] if the exponent of the power law governing the symbolic domain is sufficiently large, they will emerge almost of necessity." Synchronicity for him is not "a radical departure from the norms of nature and the otherwise ordered world of our experiences." (Hogenson 2005, p. 282) I follow his hypothesis for the levels from word association to the Self, but it seems one enters another domain when going from an intra- and inter-psychic symbolic base to the non-human outer world in mirroring that base. Since by chance the outer world rarely mirrors an inner state, synchronicity would be rarer than the occasional manifestation of the Self and thus be at the end point of a power distribution. A numinous experience of the Self would seem even more powerful if the outer world conjoined the inner. One could believe there is a God "out there" and not "just" an intra- or inter-psychic phenomenon. This argument is countered by statistical proof that a dog could repeatedly know when his owner would come home (Sheldrake 1999, p. 54-63), indicating that synchronicity is not necessarily rare or at the end of a continuum, but rather in a dimension beyond Hogenson's proposed continuum from word association to synchronicity. For a person adroit at being able to enter the Tao, a meaningful marriage of inner and outer would not be rare but would be an ongoing experience.

The DST casting of Jung's system as a power law distribution of the symbolic universe is the closest we may come to establishing a mathematical base of God. As Hogenson describes it, Jung's most overarching concept, the Self, which possesses

> the symbolic qualities of a god image, should stand at the far end of the symbolic power law distribution. And indeed, one would have to marvel at the degree to which the genuinely massive symbolic moments in human history, the emergence of the great religions, seem to possess a power of social organization that transcends anything that one would expect from a carpenter's son, a displaced prince, or the son of a minor merchant family, to acknowledge only the most recent instances of the emergence of such powerful symbolic systems. (Hogenson 2005, p. 282)

Put into an ecopsychological framework, Hogenson's hypothesis is "that the symbolic can be understood as a part of nature, sharing the characteristics of other great processes in nature, from the ion transfers in the brain to the destructive force of a great volcano." (p. 283)

V. The Human as an Embodied Robot

Applying dynamic systems theory to the human psyche makes it important to delineate the significant elements of being an embodied human. Much can be explained about archetypes by examining the basic inner and outer realities of human existence coupled with the fact that we are symbolic animals with language and story-telling abilities.

Humans are animals that walk upright, freeing our hands to manipulate the environment. As generalists, we are not adapted to unique environments yet must fill the needs for food, shelter, water and an adequate temperature range as all animals must do. We have large, programmable brains with a unique structure, functioning and development. Sexual dimorphism offers very different bases for our innate symbolizing activities to respond to both in terms of physically being in a male or female body and to vastly different hormonal experiences.

We are social creature requiring good attachment, especially in our formative early years. Our linguistic abilities allow for complex symbolic and abstracting functions and our ability to accumulate and pass on art and information are important factors enabling us to alter the face of the planet and evolve spiritually and culturally. Our sense of Self as a centering element begins with our experience of coordinated body movements. Consciousness of the end of our lives stimulates reflection on the meaning of existence, generating philosophical and spiritual concerns. These matters are considered in some detail in Appendix C: The Human as an Embodied Robot.

Humans share the basic set of emotions and affects of grief, joy, fear, anger, contempt/shame, startle and interest. Louis Stewart developed a theoretical synthesis amenable to a Dynamic Systems Theory model that illustrates how these seven affects in response to basic human conditions can give rise to *all* the dimensions of human experience. (Appendix D: Dynamic Systems Theory and Human Development)

APPENDIX C

The Human as a Situated Embodied Robot

When applying the dynamic systems model (DST) to humans (Appendix B) it is important to be cognizant of the significant elements of being an embodied human. Much can be explained about archetypes by examining the basic inner and outer realities of human existence coupled with the fact that we are symbolic animals with language and story-telling abilities. Consider the innate ability of dreams to convert our life experiences into an evolving story and how Big Dreams cast our life into a big story that rises to the level of a myth. DST indicates there will be endless and unique variations on the experience and expression of the basic themes.

One reason Jung thought all humans are basically alike is because we have a similar brain structure, but he also disavowed any attempt to link the psyche with biology. He felt we knew far too little about both realms to speculate about possible connections between the two. (von Franz 1975, p. 59; Jung 1977 p. 380-382, 466, 467) He said there are as many archetypes as there are typical life situations (CW 9, I, ¶ 99), so let's look at some of the typical life situations from the perspective of being an embodied situated human robot (Appendix B), an imagination of what it is to be in the human body and a human milieu with both in relation to the natural environment.

At the most basic level humans are living organisms in and of the natural world; we are part of nature and nature is part of us. We can live only within certain temperature ranges and we need food, water, shelter, and clean air to breath. Living in suburban homes and driving around American cities cocooned in air-conditioned cars can make these facts easy to ignore. But this ignor-ance is catching up with us as the world's food and water supplies tighten, the effects of climate change are manifesting, and air pollution continues to plague us.

We are animals meaning that we have the freedom of movement, setting the archetypal base for the American love affair with the automobile. We are animals with a wide range of sensitivities to the environment and a homothermous system maintaining an internal temperature stability which enables us to be fully conscious and active in all seasons over a wide range of temperatures. Our psyches are founded on endogenous cycles of alertness, drowsiness, hormonal changes, etc. See Appendix H: The Black Goddess for descriptions of environmental influences and sensory inputs that we are largely unaware of but unconsciously affected by, such as ionization in the atmosphere, pheromones, smells, etc.

Humans walk upright and are bipedal, giving us a heightened view of the world and a sensitive need for balance. Being bipedal frees the arms to manipulate the world with our marvelous hands with their opposable thumbs. Our eyes point straight ahead, giving us a strong directional orientation. We are very visual creatures with a huge portion of the brain devoted to optical processing, images thus assuming a dominant role in human behavior.

A significant fact is the human dimorphic—men and women are fundamentally different in many ways. The opposite sex is the same species but so different and complementary that alchemy uses sexual union as a symbol for intimate relationship between even the most basic human dichotomy—the relationship between the conscious and unconscious realms. Jung recognized the contrasexual archetype at its deepest level to be a function of the Self, where sexual union can be symbolic of union with God.

I had a meaningful workshop experience of the importance of the basic shape of the body as an aspect of sexual dimorphism. We did an exercise where we were to jump forward about three feet and when we landed we were to be a member of the opposite sex. I was surprised by how different I imagined it would be in a woman's body. My shoulders seemed much narrower. Form and function go together, meaning I wouldn't be able to throw a ball long and hard as I love to do. Playing baseball would be therefore less appealing. I imagined what it would be like to have a nice set of breasts. Being rather thinly covered, for the first time in my life I realized how women have to get accustomed to males eying them up with sexual interests in mind. This would make for very different behavior of the sexes in the dating game. I imagined my hips would be much broader and I felt greater mass there and a lower center of gravity—naturally more grounded. I would be shorter and less powerful than the average man, metaphorically "looking up" to them and a needing to be a little more cautious with them.

I couldn't imagine what it would be like to have a monthly cycling of different hormones flowing through my veins. And the thought of having a period! Bathrooms seemed much more significant. What does it feel like to be the one being penetrated during intercourse versus being in the known delightful experience of being the penetrator? I felt much more vulnerable in the receptive position, including more concern about venereal diseases and therefore greater caution about choosing a partner. And getting pregnant—drastic hormonal changes and feeling a totally different body—and something growing inside! I had a felt experience of being a real form-giver, a carrier, an encloser and container for developing, precious new life. Then giving birth after labor—how symbolic can it get! I could go on but you get the picture. All these different attitudes, perspectives, behaviors and experiences come to mind by a male human imagining what it is to be human female in form, physiology and function.

Human females come into estrus on a monthly basis and not yearly like many animals, making sexual interest uniformly high unlike most animals. This keeps the males interested and furthers connection and bonding between the sexes—and leads to many sexual games.

Hormonal differences play a huge role in feelings and behavior. Testosterone produces a more aggressive nature and a tendency to objectify. It makes the male more driven and literally "out there." Hormonal baths affect brain functioning and there are also basic differences between the wiring and structure of male and female brains. Women's brains are more balanced between the hemispheres making multitasking easier and structurally aiding greater verbal abilities.

Humans share a common repertoire of emotions and affects recognized by all cultures worldwide. These are grief, joy, anger, fear, contempt/shame, startle and interest. Louis Stewart developed a theoretical synthesis amenable to DST that illustrates how the seven basic human affects in response to basic human conditions can give rise to all the dimensions of human experience. This includes the archetypal images and stories; basic expressive dynamisms (ritual, rhythm, reason, etc.) and noetic apperception (being, becoming, etc.); the religious, aesthetic, philosophic and social cultural attitudes; imagination and exploration; the ego functions (thinking, feeling, etc.) and eros and logos. (see Appendix D: A Dynamic Systems Theory Model of Human Development)

Mae-Wan Ho describes how human consciousness, particularly the sense of self and the Self, is rooted in the archetype of the organism recognizable at all levels from the atomic to the human psychic to the galactic (see volume 1, Appendix C: Self and Organism).

Our large and complicated brains allow us to be very intelligent creatures, smart enough to reflect on the very nature of life and the universe, and to contemplate who we *really* are. We can speak, experiencing the power of the word, symbolically significant because speaking is associated with the breath of life and spirit—"In the beginning was the Word..." Words and language easily lead to abstractions and abstract thought—for better and for worse in relation to being in the body and being grounded. Our stories and myths best convey the full human experience and, together with the images and music created by artists, they can facilitate the emergence of numinous and divine moments and finding meaning in life.

Our sophisticated brains allow for reflection, rationality and a sense of an end—our inevitable death. Thoughts of death can create the ultimate anxiety or lead to a transformative experience, bringing objectivity to life and a deep appreciation of life's precious, and precarious, nature. Drawing, writing and building forms allow the creation of cultural artifacts that accumulate, expand and further develop consciousness and culture. The discovery of numbers leads to the ultimate abstractions that strangely relate to psychological and physical realities at their deepest levels, such as describing the world of sub-atomic particles and forming the numeric base of the *I Ching*. (Gardner 1974)

As conscious beings we experience and can imagine ourselves into the phases of a life cycle. Humans enter life in a totally dependent position and hopefully experience a profound love from a "good enough" mother figure. This establishes an experiential base for intimate connections not only with humans but also with the world in general. The security blanket (transitional object) is the first experience of virtual reality and it forms the basis for symbolic play, creativity and religion. (Appendix F) As social animals our need for intimate bonds leads to the creation of elaborate social, cultural, and political milieus that considerably expand our sense of being human. We are naturally inquisitive as are most animals and wonder at life and nature. Profound changes and phase shifts are experienced, especially in the first few years of life. Stable attractors for experiential and symbolizing activity go from oral and anal to the genital regions of the body. We become quite conscious intelligent beings with a sense of our selves and the world and then a revolution hits from within—puberty. We feel new "drives" and experience and literally see a change in our basic form. The connections between our emotional self and the executive functions located in the frontal lobes don't finalize until we're 21, adding to the confusion of the adolescent period. The need to earn a living in order to meet basic needs and the desire for sex and to form a family leads us into the world and jobs. Our bodies and drives weaken with age and hormonal shifts switch many male and female characteristics. Women enter menopause and are forced to identify with more than being a mother if that had been a primary focus.

A crucial element for human development that Jung recognized in the symbolism of the alchemical vessel is what contemporary psychologists call the holding environment and the analytic container. Ego consciousness needs a degree of cohesiveness to function and sustain exposure to the powers from the unconscious and the difficulties of life. The archetype of the Self provides the transpersonal base for the ego in its various stable states, while the archetype represented by Hermes and the metamorphosis of a larva into a butterfly illuminates the chaotic nature of phase transitions. A protective and sustaining container is needed for these transitional states, the metaphoric equivalent of the butterfly pupal case. This can be experienced with a real or imaginal friend or friends, an analyst and the analytic process, and/or special spots in nature, especially sacred sites that offer a sense of connection which leads to orientation and meaning.

Our symbolizing ability has many experiential bases from which to generate a sense of the Self—a virtual reality that feels like having a center and a centering energy that unites diverse elements into a meaningful whole. It begins with learning to coordinate our body movements in infancy and it has to be relearned to a lesser degree during the awkwardness of puberty as we inhabit a body of different size and form. As bipedal creatures our need for exquisite balance reinforces a sense of coordination—of meaningful interrelationship of parts and functions. Illnesses and injuries, especially back injuries, painfully bring to consciousness the importance of various body parts and their coordinated, integrated activity. Learning new physical activities requiring skills, like shooting a basketball, makes us extremely conscious of coordination and integration at physical and mental levels. The archetype of team sports is a multi-person presentation of the Self. Our social nature leads to human interactions, adding many dimensions to envi-

ronmental variation, fluctuating systems, and changing relatio
that challenge our sense of self and broaden our quest for meanin

Phases of change are fractals running across dimensions of time. T
Self extends across historical time frames, as Jung showed in his analysis of the Christian era. Images and beliefs evolve and even God can die
(Nietzsche). We see these fractals/archetypes of creation and destruction
of forms at all levels of nature from the molecular to the intergalactic,
from the succession of biotic zones to the life cycles of stars. As humans
we can consciously experience these phases physiologically and in the
virtual reality of symbolic space, with fairytales being a prime example
of symbolic space.

This is but a brief overview of the concept of the human as an
embodied robot living within many levels of the inner and outer environment. This robot offers a multifaceted and complex substantive base
as fodder to be worked on by the mysterious and beautiful creativity we
experience in dreams and the symbolizing ability of the human psyche.
We can see our psyches mirrored in the marvelous world of nature, that
inspires a consciousness to emerge that can only marvel at the vastness,
complexity, beauty—and horror—of forms and processes extending
over unimaginable dimensions of time from the fleeting existence of
subatomic particles to the life cycle of a universe.

APPENDIX D

A Dynamic Systems Model of Human Development

Jungian analyst and child psychologist Louis H. Stewart developed a comprehensive theory of psychic development based on the seven primal affects inherited by humans. Stewart's description of how the psychological functions and cultural attitudes arise fit into the framework of Dynamic Systems Theory (DST). Imagination, the symbolic play of childhood, childhood games and the stages of infant development are also explainable within that framework.

I. The Seven Innate Affective States and Their Archetypal Images

Darwin (1872) identified seven fundamental inherited emotions that have typical forms of expression easily recognizable by all humans. Grief, joy, anger, fear, contempt, shame and startle are accompanied by physiological changes peculiar to each emotion, such as the association of increased heart rate with anger. Each of the major affects is described by many different words articulating its different facets and intensities. "For example," Jungian analyst and dance therapist Joan Chodorow notes, "the expressive actions of Grief include lamenting; its intensity ranges from distress, through sadness, to the extreme of anguish." (Chodorow 1991, p. 63)

Darwin also identified many complex emotions that "are known to humans in every culture, but they lack the prototypical patterns of behavior" that painters can portray directly. These include "jealousy, envy, avarice, revenge, suspicion, deceit, slyness, guilt, vanity, conceit, ambition, pride and humility" (Chodorow 1991, p. 65, 66) that tend to be expressed in "subtle, indistinct, idiosyncratic" ways:

> However, if a complex emotion erupts into impassioned
> physical action, the clear, recognizable behavior pattern

of an innate emotion takes its place. For example, [when] guilt...shows, it seems to take on many of the expressive qualities of shame, that is, squirming, hiding, eyes averted, head hanging down.

Hatred seems to exhibit the expressive actions of Contempt as well as Rage. (p. 66)

Silvan S. Tomkins (1962, 1963, 1982) helped confirm Darwin's insights and added Interest to Darwin's list, ranging in intensity from interest to excitement. (Chodorow 1991, p. 68) Chodorow summarizes Tomkins' position:

Interest is activated by or accompanies the drives, as well as certain reflexes and functions. Interest-Excitement is an intrinsic, energic aspect of sexuality, hunger, orientation reactions, reverie and reflective problem-solving as well as looking at or listening to something 'interesting.' (Tomkins 1962, p. 339 in Chodorow 1991, p. 68)

In Tomkins' view, the affects are the primary, innate, biological motivational system of the higher mammals, including human beings. The drives and other responses are secondary...because they require amplification from the affects in order to function. For example, in a state of depression, the affect Interest is withdrawn from life's activities. Without Interest to amplify the hunger drive, we lose our appetite for food. Thus the hunger drive or the sexual drive or any other drive must be amplified by the appropriate affect if it is to work at all. Affect not only amplifies the drives, but it serves to motivate memory, perception, thought and action as well. (p. 68)

Tompkins contrasts the autonomous drive systems of simple organisms living by programmed instinct in stable environments with the developed affect systems of complex organisms living in changing environments "where survival depends on whether we can respond in flexible ways." (Tomkins 1962, p. 150 in Chodorow 1991, p. 69)

Tomkins groups affects into the positive affects of Interest and Joy, the negative affects of Distress, Fear, Anger, Contempt and Shame, and the "resetting" affect of Surprise. Louis H. Stewart developed a theoretical system of psychic development starting with Jung's thoughts in 1907 that "affects are the primary motivating system of the psyche, and...the source of imagery and consciousness." (Stewart 1986, p. 189 quoted in Chodorow 1991, p. 76) He incorporated Tomkins' hypothesis of how affects and their particular functions evolved and put Tomkins' three categories of affects into a Jungian framework. (Chodorow 1991, p. 76)

Humans respond to the perception of novelty with the positive affect of Interest, ranging from simple interest to excitement. We respond to the perception of the familiar, especially loved ones, with the positive affect of Joy, ranging from enjoyment to ecstasy. (Chodorow 1991, p. 77) Stewart saw the motivating, life enhancing affects of Joy and Interest as being "the innate, energetic source of libido, the life instinct." (p. 76) Joy and Interest potentiate each other. (Stewart 1987b, p. 133 in Chodorow 1991, p. 78) Joy expresses itself dynamically as play (imagination) and Interest expresses itself dynamically as curiosity (exploration). (Chodorow 1991, p. 78) "Play and curiosity are the early developmental forms of Jung's two kinds of thinking: fantasy thinking and directed thinking," Chodorow notes. (CW 5, ¶¶ 4-46; Stewart 1985 in Chodorow 1991, p. 79). "Curiosity is the essence of Logos or directed consciousness [a solar, differentiated, objective, definitive way of seeing]. Imagination is the essence of Eros or fantasy consciousness [a lunar, merging, subjective, imaginative way of seeing]." (Stewart 1986, p. 190-194 in Chodorow 1991, p. 150)

The five negative or crisis/survival affects

> appear to have evolved as a kind of self-protective system which sensitizes the psyche to the fundamental spiritual [existential] crises of life, namely: *loss of a loved one* (Sadness); *the unknown* (Fear); *threat to autonomy* (Anger); *rejection* (Contempt/Shame); and *the unexpected* (Surprise). (Stewart 1987b, p. 393 quoted in Chodorow 1991, p. 81)

Tompkins described the function of the "resettling" affect (Surprise) as being able to instantly reorient the system by interrupting all other affects. (Tompkins 1962, 1963; Chodorow 1991, p. 76) Stewart described this as a centering function and proposed a psychological link between the senses and the affects, suggesting "that the senses may have been the early precursors of the primal affects." (Stewart 1987c, p 393 in Chodorow 1991, p. 84)

Tompkins listed Contempt and Shame as separate affects, but both are stimulated by rejection. Stewart therefore joined the two with the determinate being whether the direction of Contempt is toward oneself or the other. (Stewart 1987b, p. 134 in Chodorow 1991, p. 77)

The primal affects tend to have accompanying images associated with pre-creation and the dawning of consciousness as found in the world's creation myths. (Chodorow 1991, p. 82, 83) Chodorow notes that the archetypal images of the *abyss*, the *void, chaos,* and *alienation* "are at the same time *experiences* of the primal affects." (p. 83) Experiencing loss in conscious life constellates the archetypal images exemplified by the *void* and accompanied by "the actual, empty feeling of Sadness." Experiencing the unknown constellates the archetypal image exemplified by the *abyss* and the feeling of falling into the "actual, bottomless grasp of Fear." The image of *chaos* appears with the sense of restriction and "the feeling of being tied up in knots, that is the terrible muddle and frustration of Anger." The conscious experience of rejection constellates an archetypal image best exemplified by *alienation* and is "the actual withering rejection we feel (toward the other, or toward one's self) in Contempt/Shame." (p. 83, 84)

Stewart suggests that the complex emotions (or feelings, affective complexes) "are mixtures, modulations and transmutations of the innate affects" that develop in the intense crucible of interpersonal relationships in the family. This generates "such subtle and complex emotions as jealousy, envy, greed, anxiety, depression, as well as respect, admiration, compassion, mercy, reverence and the like." (Stewart 1987b, p. 154 quoted in Chodorow 1991, p. 78)

Individual values (feeling is an evaluative function) and consciousness emerge out of the dynamic interplay of the seven affects. (Tompkins 1962, 1963 in Chodorow 1991, p. 68, 69) To quote Stewart, "The archetypal affects may be thought of as an innate, regulatory system of the psyche which functions as an unconscious energic, orientating and apperceptive/response system which has evolved to replace an earlier system of programmed instinct." (Stewart, 1987a, p. 40 quoted in Chodorow 1991, p. 77) Stewart's concept of the psyche as a dynamic system of interaction and complex development of the seven archetypal affects makes it easy to put archetypes into a DST framework. In DST terms, the seven innate affects are the initial programming of the situated human robot. (see Appendix B: Bootstrapping the Archetypes) They are seven distinct ways of perceiving and directly responding to the world, inner and outer, simultaneously accompanied by typical types of images. The family is the main situated environment for the process of evolving feelings/affective complexes by intermingling and transforming the affects. Affect arousal by such attractors as hunger and sex increases the dimensionality and bifurcation points in the dynamic brain-body system. This generates new and more complex systems and new information, possibilities and images. It is the creative process and it results in the growth of consciousness. Consciousness and values are developed and evolve towards wholeness only if the existential crises are fully engaged and not dissociated. For a summary of the primal affects and the symbols accompanying them see figure 1.

II. The Transformative Aspects of Play and Imagination

Stewart's sandplay therapy work with young children and his later practice as a Jungian analyst led him to the realization "that development and creativity are basically the same process." (Stewart 1986, p. 184 in Chodorow 1991, p. 71) "Play and imagination are completely intertwined with the affects," notes Chodorow. (Stewart 1977, 1978, 1981a, 1981b, 1982; Chodorow 1991, p. 72) Stewart developed the hypothesis that the affect Joy motivates and energizes play. (Stewart 1987b, p. 133 in Chodorow 1991, p. 75) It is through play that "the process of psychological development modulates and ultimately transforms the crisis/survival affects toward the development of imagery and new consciousness." (Chodorow 1991, p. 76) "Children develop a sense of who they are through symbolic play," Chodorow writes. "The symbolic play of childhood is the embodiment of the imagination...[a] play with images" (p. 72):

> The symbolic play of childhood leads children to recapitulate the emotionally charged experiences in their lives. Symbolic play involves spontaneous re-enactment of difficult situations that the child has been through. But play is completely voluntary. And no matter how difficult the content, it is fun. Here the child gets to control the situation; roles are often reversed. The process of symbolic play takes the child directly to and through the emotional core of the upsetting experience. Through playful re-enactment and further imaginative development, the overwhelming effects of the crisis emotions are modulated and transformed. (Stewart and Stewart, 1979; Chodorow 1991, p. 72, 73)

The Archetypal Affects of the Self

| Symbol and Primal Affect | | | Evolved, Differentiated Functions | |
Stimulus	Image	Affect	Expressive Dynamism / Noetic Apperception	Cultural Attitude / Ego Function
The Unknown	the Abyss	FEAR	Ritual / the Intangible	Religious Attitude / Intuitive Function
Loss	the Void	GRIEF	Rhythm / the Tangible	Aesthetic Attitude / Sensation Function
Restriction	Chaos	ANGER	Reason / Quantitative Order	Philosophic Attitude / Thinking Function
Rejection	Alienation	CONTEMPT/SHAME	Relationship / Qualitative Order	Social Attitude / Feeling Function
Unexpected	Disorientation	STARTLE	Reflection / Orientation	Psychological Attitude / Ego Consciousness
The Familiar	Illumination	JOY	Play / Being	Imagination / Eros (Relatedness)
The Novel	Insight	INTEREST	Curiosity / Becoming	Exploration / Logos (Discrimination)

Figure 1. Stewart's theoretical synthesis.
(Stewart 1987b, p. 141 in Chodorow 1991, p. 95)

Most of the problems children wrestle with are not pathological. They are basic family problems like sibling rivalry or "the essential problem of living with a mother who is sometimes good and sometimes bad." (Chodorow 1991, p. 73, 74) Stewart's work emphasizes the importance of play and imagination in child development and the individuation process of adults, confirming the significance of Hermes as the god of psychology (volume 3 of *The Dairy Farmer's Guide*). He observed that children are always playing out the emotional responses in their lives, transmuting the affects of their emotions

> through compensatory fantasies, liquidating or cathartic experiences, and the like. Thus one must conclude that during the period of development, play and fantasy serve a transformative function in the equilibration of the personality, and this makes it readily apparent why in analysis, active imagination serves an identical *transformative* function in the 're-creation' of the wholeness of personality. (Stewart 1987b, p. 133 quoted in Chodorow 1991, p. 75)

Stewart realized that "the development of the whole individual was the same process as the process we call creativity, which we envy and honor so much in the creative artist, poet, religious innovator, philosopher, and social reformer." (Stewart 1986, p. 184 quoted in Chodorow 1991, p. 71) "Play and imagination function as the dynamic source of mythic consciousness." (p. 72)

III. The Four Functions and the Origins of the Cultural Attitudes

Stewart's theoretical synthesis of Jungian psychology and contemporary affect theory also attempts to explain how the psychological functions and cultural attitudes arise out of the seven primal affects.

The apperceptive functions of the ego help us orient in the world: "*Sensation* (or sense perception) tells you that something exists, *thinking* tells you what it is, *feeling* tells you whether it is agreeable or not; and *intuition* tells you where it comes from and where it is going." (CW 18, ¶ 503) Sensation and intuition are direct forms of perception, with sensation sensing the tangible and intuition sensing the intangible world. Thinking and feeling are rational functions that order experience. "Thinking evaluates on a quantitative, logical, impersonal basis: feeling evaluates on a qualitative, organic, personal basis," writes Chodorow. (Chodorow 1991, p. 85)

Joseph Henderson (1984) described four basic attitudes that comprise the cultural unconscious: "the aesthetic, the religious, the philosophic and the social—as well as a central, emerging, self-reflective psychological attitude." (Chodorow 1991, p. 86) The four forms of cultural imagination and ways of experiencing the life of the spirit originate in the innate primal affects expressed by children. They eagerly express their fantasies in many forms (aesthetic imagination). Their fear of the unknown—things under the bed or ghosts—and the "self-comforting, repetitive actions, so similar to prayers, to ease the transition from the day world to sleep" is the foundation of the religious imagination. Children's endless questions and imaginings about the world in their search for rational explanations forms the base for philosophical imagination. Their wrestling "with feelings about being included or excluded," and their fantasies about how to get along with others is the basis for social, ethical and moral imaginations. (p. 86)

The cultural Self can be seen as an emergent phenomena from the four primal affects that "don't stop" (see Appendix B: Bootstrapping the Archetypes):

> Henderson (1984) characterizes the four main cultural attitudes: "The ethical consistency of a social attitude, the logic of a philosophical attitude, the transcendent nature of a religious attitude...[and] the sensuous irrationality of the aesthetic attitude." (Chodorow 1991, p. 49)

> The aesthetic attitude seeks the world of beauty. The religious attitude seeks the realm of the sacred. The philosophic attitude seeks principles of universal validity. The social attitude seeks a state of communal well being, utopian communitas. As we see, the cultural attitudes lead us to the age-old ideals of ultimate human value: The Beautiful, the Holy, the True and the Good. The functions of the cultural Self are mirrored in the external cultural forms of Art, Religion, Philosophy and Society. (p. 86, 87)

The ultimate value of self-knowledge and self-reflection emerges as a fifth cultural function via a convergence and quintessence of the four cultural attitudes: aesthetic, religious, philosophic and social. (p. 87)

Stewart's hypothesis is that the four primal affects have an apperceptive aspect (how one perceives things) from which emerge the ego functions and an expressive aspect that gives rise to the cultural attitudes. The dynamism of Curiosity/Exploration is particularly important for the development of the ego functions while the dynamism of Play/Imagination is important to the development of the functions of the cultural Self. (Chodorow 1991, p. 87)

The expressive behavior of sadness is *rhythmic harmony* emerging with the rhythmic rocking back and forth in deep grief. Chodorow writes:

> Sadness connects us not only to those we love, but to every aspect of the tangible world. To lose a loved one is to lose the abundance and beauty of nature...As we sense the importance of our relationship to the *tangible world*, the element earth, the ego function of *sensation*, is activated. (Chodorow 1991, p. 88)

The aesthetic attitude emerges from keeping a focus on the lost one through poetry, songs, painting and sculpture as Jung did when he lost significant people in his life. (p. 88)

The ego function of intuition emerges from fear—the sense of the "presence of myriad, *unknown* possibilities." (Chodorow 1991, p. 89) The cultural attitude of *religion* emerges from "the primal experience of terror." *Ritual* is its expressive dynamism emerging from uncontrollable repetitive action to ward off demons or appease the gods. (p. 89)

The primal affect of anger arises from a perceived threat to our survival as we experience *restriction* and a sense of *chaos*. It leads to the ego function of *thinking* as we identify the problem and try to solve it by developing *strategies* to get rid of threats and "to put things in order again." (Chodorow 1991, p. 90) The expressive behavior or cultural form is *reason*, first expressed at the primal level as threat and attack and later a symbolic attack on problems. The *philosophic* attitude arises through "the nature of the philosopher as one who thinks more deeply and obstinately than other people" when tackling unsolvable problems. (Stewart 1987b, p. 141, 142 quoted in Chodorow 1991, p. 90)

The affect Contempt/Shame arises when we experience *rejection* and we are "thrown into a state of *alienation*." The ego function of *feeling* arises as our *sensibilities* develop to "help us evaluate the subtle and complex network of human relationships." (Chodorow 1991, p. 91) Contempt/Shame is "an acute evaluative function which finds others or ourselves wanting." (Stewart 1987b, p. 153 quoted in Chodorow 1991, p. 91) Contempt/Shame leads to the development of the social attitude. "Contempt/Shame forces full attention on our place in the human community," Chodorow notes. "This punishing affect is always expressed within the context of a relationship (either with one's self or with another)." (p. 92)

These four crisis affects each have an expressive dynamism "that develops from a completely untransformed, primal state toward the higher functions of the ego and Self...Rhythm, Ritual, Reason and Relationship." (Stewart 1987b, p. 138-143 in Chodorow 1991, p. 92) Chodorow summarizes: "Sadness is expressed and transformed through rhythm and rhythmic harmony. Fear is expressed and transformed through ritual. Anger is expressed and transformed through reason, primitive and differentiated. Contempt/Shame is expressed and transformed through relationship." (p. 92) Stewart's theoretical synthesis is presented in figure 1.

IV. Games Children Play

Affects and their expressive dynamisms are seen in the universal nature of games children play. The games deliberately constellate specific emotions that are transformed in the process of playing the games. (Stewart 1987, p. 37 in Chodorow 1991, p. 153) The four categories are "games of physical *skill*, games of *chance*, games of *strategy* and *central person* games. Players have to bear, contain and ultimately transform such feelings as disappointment (physical skill), anxiety (chance), frustration (strategy), and embarrassment (central person)." (p. 153)

Chodorow reviews the four types of games. "Games of physical skill foster the development of rhythmic harmony" that comes and goes, leading to disappointment. Sensation types like these games and playing them is related to the development of the aesthetic attitude. (Chodorow 1991, p. 153)

Games of chance like throwing dice foster the development of ritual. "The outcome of the game is in the hands of the Gods" which generates fear and anxiety. Rituals develop as we repeat actions associated with a win. Intuitive types like these games and there "seems to be an archaic link between games of chance and the religious attitude." (Chodorow 1991, p. 153, 154)

"Games of strategy [like checkers] foster the development of reason," Chodorow notes. A sustained, intensely concentrated focus is developed. "Irrational impulses that come up tend to get channeled into the symbolic attack of the game." Thinking types with a tendency toward a philosophic attitude like strategy games. (Chodorow 1991, p. 154)

"Central person games [like tag and musical chairs] foster relationship...[They] deliberately evoke the experience of rejection. But the rules of the game are fair enough and the game is fun enough that such painful experiences can be tolerated—even integrated." These games help develop the feeling function and a social attitude. (Chodorow 1991, p. 154)

V. Childhood Development and the Individuation Process

Stewart also described five states of child development that naturally unfold prior to the discovery of "pretend'" and symbolic play. Development requires a good enough mother with a "capacity for spontaneous play and natural curiosity" (Chodorow 1991, p. 96):

> Each new stage of the infant's development represents an integration of a new level of mobility; a new level of consciousness; a new development in play/imagination, i.e. a new 'game'; and a new development in curiosity/exploration, i.e. a new 'interest.' In total each new stage may be described as the achievement of a new stage in the development of ego-Self identity. (p. 97)

What are called development stages can be framed in DST terms as relatively stable attractors as individuals form competencies (basins of attraction) around particular tasks (attractors). (see Appendix B: Bootstrapping the Archetypes) Contents self-organize with each phase state shift which is experienced as "a startling, numinous moment of synthesis and reorientation" as one emerges from the chaos of the transition phase. (Chodorow 1991, p. 97)

Chodorow describes the "five critical stages in normal development beginning with the first integration following birth and ending with the conscious awareness of the ability to 'pretend,' which occurs around 16 to 18 months of age. Conscious awareness of 'pretend' leads to the development of symbolic play and imagination." (Chodorow 1991, p. 97)

Stage one is uroboric wholeness, the initial union with the Self symbolized by the uroboros—the circular snake with its tail in its mouth:

> It is of at once holding and being held. Shared rhythms of holding, touching, gazing, lulling and lullabies are the psychic nourishment of this earliest phase. Parent(s) and infant immerse in these together. The infant also does them when alone. (Chodorow 1991, p. 97)

The incarnated uroboros represents an undifferentiated state of self-other consciousness with the discovery of thumb sucking being the first principle recognition of the self, the first synthesis of the psyche. (Stewart 1986, p. 191 in Chodorow 1991, p. 97, 98) This natural process of discovering "nature's pacifier" has been recognized as an important developmental event. "Not only the mouth, but the thumb too experiences the pleasurable, self calming sensations of rhythmic sucking. In this way we first learn that we are able to hold and comfort ourselves." (Chodorow 1991, p. 98)

The second stage commences with the first smile when the infant "first consciously recognizes the other." This first sign of love, this first experience of love, sometimes occurs as soon as the end of the first month "when the infant, awake and clear eyed, smiles in what is unmistakably a pleased recognition of the familiar sounds and face of the mother." (Stewart 1984, p. 1 quoted in Chodorow 1991, p. 98, 99)

Within days or weeks of the first smile comes the third stage—the first laugh that is the first recognition of the Self:

> The infant's first spontaneous laughter when alone express-es joy in the sheer exuberance of bodily motion [such as leg-kicking]...Or, the baby throws his or her head back and discovers that s/he is both the mover and the one who is moved...Natural laughter has at its core 'the spontane-ous expression of the pure joy of being alive and ...is...the prototype of play' (Stewart 1985, p. 93). From this point on, laughter will mark every new recognition of the self. (Chodorow 1991, p. 99)

The fourth stage is the establishment of object constancy—the things of the world are there even if not always seen. It develops around issues of separation anxiety in the third quarter of the baby's first year. Peek-a-boo and later games of hide-and-seek help the infant deal with the pain of separating from beloved parents and gradually establish object constancy. (Chodorow 1991, p. 99, 100)

The fifth stage occurs at about 16 to 18 months of age when the baby discovers s/he can pretend. (Chodorow 1991, p. 100) The child often discovers non-verbal, symbol-creating (semiotic) activity through miming

> an already adaptive behavior pattern like the ritual behavior adopted to ease the transition into sleep (e.g., thumb sucking and fingering the satiny edge of a blanket). The child laughs with joy at this new recognition of Self; and this is pretend play (Piaget 1962)...Pretend play begins with the miming of a behavior pattern which has assisted the child in warding off fear of the unknown and soothing the anguish of separation. Subsequent pretend play will be seen to reenact all of the emotionally charged experiences of the child's life. (Stewart and Stewart 1979, p. 47 quoted in Chodorow 1991, p. 101)

The fact that imagination is now conscious is demonstrated by the child teaching the sleep ritual to all the dolls. This passage into the symbolic world coincides "with the beginning of real curiosity about language," another dimension of the symbolic. Each of these stages is associated with a particular quality of movement that appears in dance therapy. (p. 101)

Developing into a whole person is part of the individuation process with the energy for the process arising from the conflict between opposites. The archetypal affects with their accompanying images automatically constellate their opposites. The resulting tension, conflict and discord are uncomfortable but serve the purpose of a development towards wholeness. "The process simultaneously leads to the realization of the Self"—a complex of opposites—the psyche. (Chodorow 1991, p. 92, 93)

Chodorow lists the primal images and their compensations:

> The emptiness of the *void* [of Sadness] is compensated by the abundant *beauty of nature*. The *abyss* [of Terror] is compensated by the *holy mountain*. *Chaos* [of Rage] is compensated by the *ordered cosmos*. *Alienation* [of Contempt/Shame] is compensated by *utopian communitas*. (Chodorow 1991, p. 93)

These images pairs, with one image in consciousness and the other in the unconscious, "span the heights and depths of the human emotional experience." (Chodorow 1991, p. 93) Stewart sees in these pairs a spiritual dimension "indicative of the highest of human values...the age old categories of the holy, the beautiful, the true and the good, which have everywhere found expression in the cultural forms of religion, art, philosophy and society." (Stewart 1987a, p. 43 quoted in Chodorow 1991, p. 93)

APPENDIX E

Hermes as God of Dynamic Systems Theory

Many Greeks knew the gods didn't literally exist yet they experienced powerful forces in themselves and in the world. Personifying these forces in the forms of the gods and goddesses and bringing them to life with myths, images and songs helped delineate the powers and raise them to conscious awareness. (Hillman 1975, p. 13-17) The powers the Greeks experienced that are now being described by complexity theory (CT) and dynamic systems theory (DST) are largely within the domain of the god Hermes.

As god of the myth-making activity of the human psyche, what Jung called the "religious function," Hermes is god of the concept of major attractors in CT and DST. Myths ("other people's religions") are the most significant factors in determining the ethics, morals, attitudes and behaviors in a culture. Consider how the mythic image of the American cowboy stokes our self-image of being rugged individualists. Its corollary is an aversion to socialist thought, with a propensity to see it in its extreme form, communism. This makes it difficult to promote an agenda like a national health plan for America.

Hermes' consciousness of core structures and their sources comes from being a follower of Mnemosyne, the ancient Greek personification of memory. She communicated through evocative forms like stories, myths, dreams and thoughts that stimulated musings—she was the mother of the Muses.

Myths as attractors produce fractals—the same basic pattern seen across all levels. Individuals can live mythic lives and cultures are determined by the myths they live by. Hermes was the messenger and translator between the gods and between the gods and humans. There are endless variations on mythic themes while myths are simultaneously what was, is, and will always be. The gods and goddesses described by myths function like fractals represented as powerful attractors in the guise of inclusive worldviews and basic metaphors. Everything fits together at all levels from the particular perspective of a god or goddess, the self-organizing element being the mythic story that enables one to see everything through the life and eyes of that god or goddess. The fractal is also the "just so," eternal nature about a god or goddess that cannot be reduced—it is a given, a personification of an essence of an aspect of reality.

Hermes is most associated with the phase transition in CT and DST by being god of the journey and not the goal. His shrines were stone piles at crossroads placed there by travelers for their protection while journeying. Uncertainties, non-linear events, and non-deterministic phenomena are associated with journeying. This element is conveyed in the *Homeric Hymn to Hermes* by Hermes' sandals that were "indescribable, unimaginable," enabling him to bound all over the place, and Hermes' trick of appearing to walk backwards when he was really going forward. "The best laid plans [goals] of mice and men often go awry"— Hermes is that unpredictable, tricky element in life and in the universe associated with thievery, the lottery, and sudden gain or loss.

The proper attitude to have while journeying is reflected in the Indian spiritual classic, *The Bhagavad-Gita*. Krishna counsels Arjuna, "You have control over action alone, never over its fruits. Live not for the fruits of action, nor attach yourself to inaction." (Yogi 1967, 11:47 quoted in Abraham 1995, p. 65) Arjuna is "not to worry about the consequences of battle; the importance is in the act of self-affirmation: 'Slain you will reach heaven; victorious you will enjoy the earth. Therefore...stand up resolved to fight.'" (Yogi 1967 quoted in Abraham 1995, p. 69)

Hermes is god of hinges and pivot points, dawn and twilight, spring and fall, binding and unbinding, and initiations—the "betwixt and between" space of transitional phenomena in DST. His association with unpredictability and turning points are related to the DST concept of sensitive dependence on initial conditions.

Hermes has strong associations with emergence phenomena in the phase transitions of DST. He is symbolic of the *source* of the sun in a ram-carrying ritual. (Kerenyi 1976, p. 85-87) As bearer of the sun Hermes is related to the creativity of the gap, the bifurcation point (see Appendix A: Dynamic Systems Theory) as presented in

> the kabbalistic doctrine of the *eyin*, the great Nothing...the *ein sof*, the infinite potential of all existence. All creation had to pass through this stage of nothingness. This was the moment of contact with the *sefira of chochma*, the

profoundest level of creation that precedes all differentiation and is therefore beyond comprehension. Hence the
moment of temporary turning of something into nothing
had to precede every great moment of self-renewal. (Weiner
1969 quoted in Abraham 1995, p. 69)

Frederick Abraham describes the word *tohu* in the Kabbalah as "a
state of undefined mystic creativity from which a healing bifurcation,
a *tikkun*, brings order" after experiencing the anxiety of being in the
unstable bifurcation point. (Abraham 1995, p. 69)

As old man and young male alike, Hermes represents both sides of a
transition phase. Swift-as-death Hermes and his ability to volatilize and
become invisible is related to the breakdown of old forms (the father or
the old king in alchemical imagery). Old forms get restructured, revital-
ized and reborn into the son who is symbolic of emergent phenomenon. The ability to create new psychological states and new products
is an aspect of Hermes' virginal nature, with the results symbolized by
the divine child. The child is innocent and vulnerable, yet powerful
and threatening to the powers that be: day-old Hermes was frightening
to Apollo in the *Homeric Hymn* and Herod killed off the newborn males
after Jesus' birth.

DST phenomena are symbolized by Hermes' wand. Polarities,
conflicts and the effects of competing strange attractors are represented
by the arms of the wand. The arms can symbolize different or opposing
attractors relating to each other, opposites like conscious and unconscious, predators and prey, etc. Interactions that can lead to transitional
phenomena occur in the gap in the wand, with both sides being transformed in the transitional phase symbolized by the gap. The narrower
the gap in the wand—the interactive field from which a bifurcation
point can arise—the more intense the interaction and the higher the
dimension/complexity in the dynamic system. This can be envisioned
in human relations as the degree of intimacy (closeness), excitement,
creativity and spontaneity that arises when lovers interact. The intoxi-
cating physiological dimensions of close human contact are discussed
in Appendix H: The Black Goddess. The greater the fascination with
the other, the more one is drawn into intimacy/close relationship. A
sense of "I and Other as One" can emerge, uniting inner and outer
in a mysterious *conjunctio*. Winnicott's emphasis on the mother-infant
bond, the fundamental example of attachment theory upon which
modern psychoanalytic theory is based, also emphasizes the creative,
bifurcating gap. The mother is "beyond herself" in her enthrallment

with her newborn, and her devotion spurs in her infant the creation of nothing less significant than a new human psyche. Winnicott emphasized the "tremendous significance" of the interplay of two objects in close contact like the edges of two curtains. (Winnicott 1966, p. 369)

Hermes relationship to DST is further conveyed by the professions for which he is god: diplomats, businessmen, psychologists, translators and ecopsychologists deal with complex systems and the subtle interactions and shifts within and between them.

Hermes' association with archetypal trickster figures is related to "the edge of chaos" concept. Tricksters were considered to be cultural heroes and world creators who inspire and inject creative energy. They, like Hermes, are associated with good and bad luck, often with apparent misfortune that turns out to be good luck, spoiling plans as they punish pride, arrogance, and insolence. Tricksters also prevent premature closure which limits creative possibilities. Trickster energy pushes the psyche to the edge of chaos which can lead to expansion and the creation of new forms. Joanne Lauck notes, "Although its lack of concern for our fears, the culture's taboos, or social appropriateness is unnerving and can feel punitive, the Trickster's demands for a change of direction or for stillness is a call for a necessary change of some kind." (Lauck 2002, p. 187)

Loss of old ego positions and worldviews feels like death as often metaphorically depicted in dreams. It takes courage to face life's difficulties and oppositions. Abraham notes that the great theologian Paul Tillich wrote about "the courage to confront existential anxiety" in which "'new meaning is created and recognized'" out of "the dynamics of the interaction of inner and outer realities" (Abraham 1995, p. 70):

> Life is ambiguous.
> Courage is the power of life to affirm itself in
> spite of this ambiguity.
> (Tillich 1952 quoted in Abraham 1995, p. 65)

The struggle to survive can raise psychic energy, thereby increasing dimensionality (see Appendix A: Dynamic Systems Theory). Necessity being the mother of invention is mythically portrayed in the story of Eros who shares many traits with Hermes. (Kerenyi 1976, p. 55, 56)

Nietzsche poetically addressed the "enantiodromic bifurcations arising from the dynamics of dialectical opposites" (Abraham 1995, p. 68):

A dangerous crossing, a dangerous wayfaring, a dangerous
looking-back, a dangerous trembling and halting.
What is great in man is that he is a bridge and not a
goal...

Man is something to be surpassed.
Man is stretched between the animal and the
Superman—a rope over an abyss.
(Nietzsche 1928 quoted in Abraham 1995, p. 65, 69)

Nietzsche railed against complacency (old attractors), urging us to
strive to become the Superman—to affirm life by going beyond ourselves
through confrontation, development and "surrender as a part of the
courage of will, of crossing the gap." (Abraham 1995, p. 69)

APPENDIX F

Winnicott's Transitional Object

The British psychoanalyst Donald W. Winnicott located the origin of religion and the cultural experience in the baby's experience of the first not-me object, what he called the *transitional object*. This in-between phenomena is Hermes' domain, with cultural phenomena arising from the experience of a transitional object that "doesn't stop." (Appendix B)

Winnicott explores the phase between the newborn's first auto-erotic activities, like fist-in-the mouth, and the eventual attachment to a "security blanket." This is the first possession in "the intermediate area between the subjective and that which is objectively perceived." (Winnicott 1951, p. 231) *Transitional phenomena* commence between 4-12 months with infants caressing their face without thumb sucking or sucking their thumb while the other hand caresses the face. The free hand may be used to take an external object, like a part of a sheet or blanket, into the mouth along with the fingers and suck or not suck this piece of cloth. The baby may also use a soft object to caress itself (p. 231) or engage in mouthing accompanied by babbling sounds. (p. 232) A *transitional object* may emerge out of these experiences, like a corner of a blanket or a soft object carried around while traveling. It may also be a word, tune, or mannerism, "which becomes vitally important to the infant for use at the time of going to sleep, and is a defense against anxiety" (p. 232):

> The infant assumes rights over the object...[that] is affec-
> tionately cuddled as well as excitedly loved and mutilated...
> It must never change, unless changed by the infant...It
> must seem to the infant to give warmth, or to move...or to
> do something that seems to show it has vitality or reality
> of its own...It comes from without from our point of view,
> but not so from the point of view of the baby. Neither does
> it come from within; it is not an hallucination...Its fate is
> to be gradually allowed to be decathected...In the course of

> years...it loses meaning, and this is because the transitional phenomena have become diffused, have become spread out over the whole intermediate territory between 'inner psychic reality' and 'the external world as perceived by two persons in common,' that is to say, over the whole cultural field. (p. 233)

"It is true that the piece of blanket (or whatever it is)," Winnicott notes, "is symbolical of some part-object, such as a breast. Nevertheless the point of it is not its symbolic value so much as its actuality. Its not being the breast (or the mother) is as important as the fact that it stands for the breast (or mother)." Use of a transitional object is part of the process of being able to accept difference and similarity; it describes "the infants journey from the purely subjective to objectivity." (Winnicott 1951, p. 233, 234) "When symbolism is employed the infant is already clearly distinguishing between fantasy and fact, between inner objects and external objects, between primary creativity and perception." (p. 233) "The transitional object is *not an internal object* (which is a mental concept)—it is a possession. Yet it is not (for the infant) an external object either...The transitional object is never under magical control like the internal object, nor is it outside control as the real mother is." (p. 237) "*The intermediate area...is the area that is allowed to the infant between primary creativity and objective perception based on reality-testing. The transitional phenomena represent the early stages of the use of illusion.*" (p. 239) "This early stage in development is made possible by the mother's special capacity for making adaptation to the needs of her infant, thus allowing the infant the illusion that what the infant creates really exists." (p. 242) Reality acceptance is never completed, and the strain of relating inner and outer reality is relieved by the intermediate area of experience of the arts, religion, etc. (p. 240)

The child's employment of a transitional object, the first not-me possession, is its first use of a symbol and first experience of play. (Winnicott 1966, p. 369) "*Play is...neither a matter of inner psychic reality nor a matter of external reality*" (p. 368):

> The object is a symbol of the union of the baby and the mother (or part of the mother). This symbol can be located. It is at the place in space and time where and when the mother is in transition from being (in the baby's mind) merged in with the infant and being experienced as an object to be perceived rather than conceived of. The use of an object symbolizes the union of two now separate things,

baby and mother, *at the point of the initiation of their state of separateness.* (p. 369)

"Cultural experiences...provide the continuity in the human race which transcends personal existence. I am assuming," Winnicott writes, "that cultural experiences are in direct continuity with play, the play of those who have not yet heard of games. " (Winnicott 1966, p. 370) "Cultural experience is located...in the *potential space* between the individual and the environment (originally the object)." (p. 370, 371) Symbolically this is the gap in Hermes' wand. This potential space is sacred to the individual because "it is here that the individual experiences creative living" (p. 372):

> In the average good experience in this field of management...the baby finds intense, even agonizing, pleasure associated with imaginative play. There is no set game, so everything is creative, and although playing is part of object-relating whatever happens is personal to the baby. Everything physical is imaginatively elaborated, is invested with a first-time-ever quality. Can I say that this is the meaning intended for the word "cathect"? (p. 371)

To an observer everything a baby does has been done and felt before and the transitional objects have been adopted and not created:

> Yet *for the baby* (if the mother can supply the right conditions) every detail of the baby's life is an example of creative living. Every object is a "found" object. [Hermes] Given the chance, the baby begins to live creatively and to use actual objects to be creative into. If the baby is not given this chance then there is no area in which the baby may have play, or may have cultural experience; then there is no link with the cultural inheritance, and there will be no contribution to the cultural pool.

> The "deprived child" is notably restless and unable to play, and has an impoverishment of capacity to experience in the cultural field...Failure of dependability or loss of object means to the child a loss of the play area, and loss of meaningful symbol. In favorable circumstances the potential space becomes filled with the products of the baby's own imagination. In unfavorable circumstances the creative use of objects is missing or relatively uncertain. I have described elsewhere the way in which the defense of the compliant false self appears, with the hiding of the

true self that has the potential for creative use of objects.
(Winnicott 1966, p. 371)

Winnicott quotes from a paper by Plaut (1966), a Jungian analyst: "The capacity to form images and to use these constructively by recombination into new patterns is—unlike dreams or fantasies—dependent on the individual's ability to trust." Winnicott comments: "The word *trust* in this context shows an understanding of what I mean by the building up of confidence based on experience, in the area of maximal dependence, before the enjoyment and employment of separation and of independence." (Winnicott 1966, p. 372) Psychologists now incorporate the word "trust" into the concept of "secure attachment."

APPENDIX G

The Sacred Prostitute and the Erotic Feminine

The archetype of the sacred prostitute offers a stark contrast to the Judeo-Christian image of a positive feminine figure. Two Jungian analysts, Nancy Qualls-Corbett in *The Sacred Prostitute: Eternal Aspect of the Feminine* and Rachel Hillel in *The Redemption of the Feminine Erotic Soul*, provide a rich investigation of this intriguing and important archetype. Examining the archetype from the perspective of dynamic systems theory can describe how the experience of Divine love can emerge out of erotic experiences, moving one beyond an egotistical focus and into a sensuous relationship with the body and with nature.

I. History

Sacred prostitution existed within the ancient matriarchal systems, contrasted with the patriarchy by custom versus law, religious authority versus military power, tradition-bound cohesion of the collective versus the worship of the *aresteia* of the individual warrior, and cultural authority versus political power. (Thompson 1981, p. 149, 150 in Qualls-Corbett 1988, p. 30) The gods and goddesses in these ancient matriarchies were nature and fertility divinities because people lived close to nature. The deities determined destiny by "providing or denying abundance to the earth" and individuals. (p. 30, 31)

The Love Goddess was one among many archetypal representations of ancient female deities. The goddess of love, passion and fertility was known as Inanna to Sumerians, Ishtar to Babylonians, Anahita to Persians and Astarte or Anath to the Canaanites, Hebrews and Phoenicians. She was Cybele in Lydia and Isis, earlier identified with Hathor, in Egypt. (Qualls-Corbett 1988, p. 57) In Greece she was called Aphrodite and in Rome she was Venus. She was a dynamic transformative principle who blessed her ritual participants with experience of the sublime through erotic-sensual passion. (see Appendix E: The Black Goddess for a physiological explanation of how this is possible.) "Surrendering to sensual-sexual desires was seen as a devotional path, a religious channel to worshipping the Love Goddess," notes Hillel. (Hillel 1997, p. 102)

One can get a sense of the goddess through an ancient Sumerian account (ca. 3000 BCE) of their most important celebration; a multi-day New Year festival during the summer solstice that was part of Inanna worship. During the celebration the temple of love as "the locus of potency and fertility" was where great feasts were held and ample amounts of beer and wine had been prepared for the festival. Sacrifices of first grains and fruits and offspring of livestock were made in thank-fulness to the goddess. Lively music from temple musicians encouraged dancing and love-making. (Qualls-Corbett 1988, p. 24) The baccha-nalian dance of the sacred prostitute, not unlike the ecstatic dancing maenads who followed Dionysus, connected participants to profundity of spirit through erotic ecstasy—wisdom achieved by going beyond the conventional modalities of being. "The dance was a madness filled with prophesy and secret knowledge," (Otto 1981, p. 144 quoted in Qualls-Corbett 1988, p. 70) archetypically related to Hermes' ecstatic bees. The celebration culminated in the *hieros gamos*, the sacred marriage,

> [a ritual] reenactment of the marriage of the goddess of love and fertility with her lover, the young, virile vegeta-tion god. The chosen sacred prostitute, a special votary who is regarded as the personification of the goddess, unites with the reigning monarch, identified with the god. This union assures productivity of the land and fruitful-ness of the womb of both human and beast; it is the "fixing of destinies." (Hooke 1963, p. 54 quoted in Qualls-Corbett 1988, p. 24)

After much feasting and merriment and buoyed up by ecstatic love songs, the couple retired to a perfumed nuptial bed in the sacred chamber of the temple tower. The populace serenaded the couple with hymns and love songs to enhance their rapture and fertilizing power. (Qualls-Corbett 1988, p. 24, 25)

Inanna, a prominent deity in the Sumerian pantheon, was Queen of Heaven and Earth, the Morning Star and Evening Star. Referred to as "you who sweeten all things," she brought to earth the gifts of civi-lization and culture, including the art of love-making. (Qualls-Corbett 1988, p. 32) Inanna was called "A Sacred Priestess" as well as "A Hierod-ule of Heaven" (Hillel 1997, p. 118), hierodule being a commonly used description of the sacred prostitute meaning "sacred servant." (Qualls-Corbett 1988, p. 25 note 9)

Ishtar was the goddess of love, passion, war and death in ancient Babylonia. She was "the sweet voiced mistress of the gods" and was "known for her cruel and relentless fickleness toward her lovers." This tempting, full-breasted goddess had the power to give or withhold love and sexual joy, thereby exercising control over the life cycle. (Qualls-Corbett 1988, p. 32, 33) She was called the Great Goddess Har, Mother of Harlots, and her high priestess was considered the spiritual ruler of "the city of Ishtar." (Walker 1983, p. 820 in Qualls-Corbett 1988, p. 33)

The sacred prostitute on earth was to embody and thereby serve the goddess. She was schooled in the art of love-making and served as the vessel to bring sexual joy that transformed the raw animal instincts into love and love-making. (Qualls-Corbett 1988, p. 34) Her image "represents the vital, full-bodied nature of the feminine." (p. 14) The beauty and sensuality of the sacred prostitute were seen as gifts bestowed by the goddess to be expressed in a graceful manner. (p. 29, 30) Sacred prostitutes in later civilizations

> were often known as Charites or Graces, since they dealt in the unique combination of beauty and kindness called *charis* (Latin, *caritas*) that was later translated 'charity.' Actually it was like Hindu *karuna*, a combination of mother-love, tenderness, comfort, mystical enlightenment and sex. (Walker 1983, p. 820 quoted in Qualls-Corbett 1988, p. 34)

There were several types of sacred prostitution in the ancient world. According to Babylonian law, every woman had to serve at least once in her life in the temple of love. She did this before marriage as a sacred initiation into womanhood, accepting any man who entered the temple and chose her. The stranger was viewed as an emissary of the gods whose love-making and offering to the goddess were considered to be holy acts. (Goldberg 1930, p. 78 in Qualls-Corbett 1988, p. 35) The woman returned home, blessed and honored, "usually to prepare for her forthcoming marriage." (p. 34, 35)

In some countries only the highest born could be sacred prostitutes. The most aristocratic Roman matrons engaged in sacred prostitution in the temple of Juno Sospita when the country needed a revelation. (Walker 1983, p. 820 in Qualls-Corbett 1988, p. 36) The Vestal Virgins of Greece and Rome spent their entire lives in the temple (Walker 1983, p. 821 in Qualls-Corbett 1988, p. 36) whereas great numbers of temporary hierodules frolicked in the streets during the great festivals of the fertility goddesses in Egypt. (Hastings 1956, vol. 6, p. 676 in Qualls-Corbett 1988, p. 36) There were over a thousand sacred prostitutes at the temples of Aphrodite in Eryx and Corinth and about six thousand in residence at each of the two Comanas. (Hastings 1956, vol. 6, p. 675 in Qualls-Corbett 1988, p. 37) They were educated, accorded social status and in some cases were politically and legally equal to men. (Qualls-Corbett 1988, p. 37)

II. The Archetypes of the Love Goddess and the Sacred Prostitute

Qualls-Corbett noted several features of the sacred prostitute that are important for raising the consciousness of modern women. Her body, beauty and sensuality were devoted to the worship of the goddess of love and not used to gain power, possessions, or security. Her identity was rooted in her own womanliness reinforced by her devotion to the divine feminine and did not require a male for her sense of self. Because she often remained anonymous and veiled, love-making was not for attaining a male's adoration or devotion. She was the holy vessel through which the divine feminine was incarnated in love making, thus unifying the chthonic and the spiritual. Human emotions and creative bodily energies were transcended and united with the suprapersonal. By embodying basic and regenerative energies in a divine manner, she "assured the continuity of life and love." (Qualls-Corbett 1988, p. 40)

"Sexuality is the strongest symbol at the disposal of the Psyche," Jung proclaimed. (C. G. Jung, transmitted by Dr. Eleanor Bertine quoted in Hillel 1997, p. 114) "Sexuality is not mere instinctuality; it is an indispensable creative power that is not only the basic cause of our individual lives, but a very serious factor in our psychic life as well." (CW 8, ¶ 107) To quote the poet Rilke: "Artistic creation lies so incredibly close to sex, its pain and its ecstasy, that the two manifestations are indeed but different forms of one and the same yearning." (Rilke 1934, p. 30, 31 quoted in Hillel 1997, p. 116)

Theologian and Jungian analyst Ann Ulanov commented, "The highest phase of confrontation and individuation in both sexes is initiated by the feminine," indirectly for a man through his anima and directly for a woman through the feminine self. "The feminine...completes the individuation of *each* sex. The masculine initiates the emergence of consciousness from primary unconsciousness; the feminine initiates the completion of consciousness by re-establishing contact with the unconscious." (Ulanov 1971, p. 269 quoted in Qualls-Corbett 1988, p. 56) (Hermes is symbolic of the activities of the unconscious and its relationship to consciousness, involved in the development of both masculine and feminine consciousness.).

There are dual aspects of the feminine. A maternal aspect is elementary or static: "It is the unchanging and stable factor that fosters feelings of security, protection and acceptance." (Qualls-Corbett 1988, p. 56) A transformative aspect "accents the dynamic elements of the psyche that urge change and transformation." It can be like a "divine madness of the soul"

> which invokes primeval forces that take us out of the limitations and conventions of social norms and the reasonable life. Eros in this sense produces ecstasy, a liberation from the conventions of the group...Ecstasy may range from a momentary being taken out of oneself to a profound enlargement of personality. (Ulanov 1971, p. 159 quoted in Qualls-Corbett 1988, p. 56)

Transformative energy is associated with the love goddess and her embodiment as the sacred prostitute. Creativity is stimulated, new attitudes arise, and life becomes exciting and meaningful—like being in love. (Qualls-Corbett 1988, p. 57) (see John Haule, 2010, *Divine Madness*)

Archetypal images of the sacred prostitute still emerge in the psyche accompanied by a dynamism that changes our consciousness through a shift in emotions, attitudes and ideas. (Qualls-Corbett 1988, p. 57) Archetypes manifest via symbols and "For those having the symbol, the transition is more easily made." (An old alchemical saying quoted in Hillel 1997, p. 67) The experience of the constellated archetype of the sacred prostitute is to be emotionally energized and have life "imbued with the vitality of love, beauty, sexual passion and spiritual renewal," writes Qualls-Corbett. (Qualls-Corbett 1988, p. 14) Without her "there is a certain barrenness to life. Creativity and personal development are stifled...[and] the needs for relatedness, feeling, caring or attending to nature go unheeded." (p. 14-16) Marion Woodman notes the dark side in her absence: "So long as we are unconscious of the divinity inherent in matter, sexuality can be manipulated to fulfill ego desire...The Goddess is called upon to justify lust and sexual license." (p. 9)

The Roman Venus, derived from the Greek Aphrodite, made an important separation of love and passion from fertility—that was Rhea or Demeter's realm. Aphrodite was also known for her laughter, part of an interesting gestalt of traits dissociated from the more purely biological drive for reproduction. (Qualls-Corbett 1988, p. 57, 58)

The love goddesses were associated with springtime, blooming and nature splendidly bursting forth from dormant seeds. (Qualls-Corbett 1988, p. 57, 58) "Beauty is the quintessential component" of the love goddesses and lovely smells are important. The loveliness of Aphrodite's body was adorned and adored. Her nakedness was glorified; she was "the only goddess to be portrayed nude in classical sculptures." (p. 58)

Aphrodite ornamented herself in gold and "was often referred to as "the golden one." (Qualls-Corbett 1988, p. 58) Goldenness symbolized her radiance and freedom from pollution, gold being a non-corrosive element. It also denotes a type of consciousness that mythologist Kerenyi describes as "[not] heavy or darkly earthly...but...something bright and lucid." (Kerenyi quoted in Qualls-Corbett 1988, p. 58) It is the consciousness of relationship and feeling, "not spiritual but earthly consciousness 'exalted to the highest purity.'" (Bachofen quoted in Qualls-Corbett 1988, p. 58) Qualls-Corbett notes how the goddess exemplified "those aspects of feminine nature which are manifested in matter." Her traits include "physical beauty, feminine consciousness integrated in the body (i.e., instinctive wisdom) and the interconnecting capacity for deep-felt emotions and relatedness (the principle of Eros)." (Qualls-Corbett 1988, p. 58)

Aphrodite was a virgin despite having many lovers, virginal in the sense of belonging to herself and to no man. If married the husband was viewed as a consort and wifeliness did not alter her or lend her special status. She was not seen as the feminine version of a god or as a counterpart to other gods. (Qualls-Corbett 1988, p. 58, 59) "The goddess of love exists in her own right," true to her own nature like nature, as "one-in-herself." (Harding 1971, p. 124 in Qualls-Corbett 1988, p. 59) Woman's virginal nature was described by Philo of Alexandria, a first-century Jewish philosopher:

> For it is fitting God should converse with an undefiled, an untouched and pure nature, with her who in very truth

> is *the* Virgin, in fashion very different from ours. For the congress of men for the procreation of children makes virgins women. But when God begins to associate with the soul, He brings it to pass that she who was formerly woman becomes virgin again. (Harding 1971, p. 187 quoted in Qualls-Corbett 1988, p. 63)

This golden aspect of the virginal, associated with the Divine, creativity, renewal and the emergence of the soul is the archetypal base of our need for wilderness experiences in nature.

Aphrodite was also a moon goddess because in some hot, arid environs where she was worshipped the sun was associated with the drying and killing of vegetation while the moon was associated with fertilizing powers that fostered life. The moon is also associated with a cyclic rhythm of constant change. It could bring lunacy and in its dark phase: "The goddess was ominous in her boundless rage and ruthless destruction." (Qualls-Corbett 1988, p. 59) Qualls-Corbett believes a woman conscious of the moon goddess intuitively moves with the cyclic ebb and flow of her changing energy or moods. (p. 63)

III. The Son-Lover

The son-lover is an important part of the love goddess' archetypal gestalt:

> The goddess herself is eternal, yet the son-lover is slain or sacrificed to be resurrected again. Inanna's young love was the shepherd Dumuzi, who was sacrificed to the Nether World for six months every year, as was Ishtar's son-lover, Tammuz. In Egypt, there were Isis and Osiris, in Lydia, Cybele and Attis. The theme is repeated as each young man meets an untimely, cruel death, and eventually is brought to earth or life once again. (Qualls-Corbett 1988, p. 59)

A Greek myth of Aphrodite and her beautiful lover Adonis describes the depth of pain and emotion the love goddess experiences at the death of her lover. The purified Christian version is the pieta scene of Mary holding the crucified Jesus:

> Loss and death, unrequited love and abandonment, are all part of Aphrodite's realm...Permanence is of Hera's world, not Aphrodite's. What belongs to her is a deep acceptance

> that passionate love does not last forever; and an equally
> deep acceptance that man is made to love. (Stassinopoulos
> and Beny, 1983, p. 83 quoted in Qualls-Corbett 1988, p.
> 60)

All love goddess myths emphasize this experience. "We know the range of this goddess' emotions—joy and pleasure, yet also pain and grief—to a greater extent than those of all other goddesses. Emotions engendered by love's process are an integral part of her being," writes Qualls-Corbett. (p. 60, 62)

The sacrifice of the son-lover is a challenging aspect of the love goddess gestalt. The son or significant male in a woman's life is not sacrificed for the transformation of both if the woman lives vicariously through the male or uses her maternal nature to make men dependent on her. "The strength of the goddess lies in the capacity to give up that which is most precious, in order to ensure growth and regeneration; transformation can only take place when old attitudes and values give way to new ones." (Qualls-Corbett 1988, p. 65) Qualls-Corbett continues, "Mourning is a way of consciously integrating the fact that circumstances have changed." (p. 64) Only when a loss is fully felt can one move on (described in hexagram 29, the Abysmal in the *I Ching*).

A male who remains dependent on women for security, acceptance and nurture remains fixated or regressed to the Oedipal stage (Qualls-Corbett 1988, p. 65, 66) with no responsibility or accountability. (Hillel 1997, p. 26) He is unable to establish a mature relationship with a woman. The woman remains an object for him to immediately gratify his sexual desires on demand and "the spiritual dimension of the sexual act is never experienced." The ephemeral gratification soon passes, the heart is not touched, the soul is not moved; Aphrodite's fullness as erotic *and* spiritual is missed. (Qualls-Corbett 1988, p. 66)

Eric Neumann noted:

> the anima is "the vehicle par excellence of the transforma-
> tive character" in man. "It is the mover, the instigator of
> change, whose fascination drives, lures, and encourages
> the male to all the adventures of the soul and spirit, of
> action and creation in the inner and the outward world."
> (Neumann 1955, p. 33 quoted in Qualls-Corbett 1988, p.
> 66, 67)

The archetype of the love goddess is powerful:

> She is the active principle of Eros which enables us to be related to our own emotions, and also to touch the emotional substance of another. (Qualls-Corbett 1988, p. 68)

> Aphrodite's essence is transformation through the power of beauty and love—the [psychic] power that is responsible for all metamorphoses. (Stassinopoulos and Beny 1983, p. 83 quoted in Qualls-Corbett 1988, p. 68)

For a woman the son-lover aspect of the archetype represents the spirited, adventuresome hero who rescues her from her tie to the father's conventional attitudes, bringing her creative thoughts and new attitudes. (Qualls-Corbett 1988, p. 72) A woman must withdraw the projections onto a man of her creative abilities in the personified form of the son lover. Hillel sees a Western framework that trains women to remain maidens; to develop a polite persona and "follow norms of being 'a good daughter to the Father'" maintained through a laundry list of "shoulds" and "oughts." (Hillel 1997, p. 86)

The stranger in the love goddess gestalt/archetypal image was "viewed as an emissary of the gods, or even the god in disguise." (Qualls-Corbett 1988, p. 74) In dreams, myths and fairytales he represents a breakthrough from the unconscious to instigate change. The numinosity of the divine is experienced as something "other" enters consciousness with a feeling of strangeness. (p. 75) The positive animus functions as a bridge between the woman's ego and her own creative resources. It leads her "out into the world of objects, creativity and ideas," says Qualls-Corbett. "It is that psychic function that enables a sense of direction, focus, discernment and ordered continuity in all endeavors." (p. 76)

The animus can appear positively as a stranger, wise old man, a youthful Adonis, or baby boy; negatively as a rapist or robber "who takes the woman's most prized possessions, symbolic of her feminine nature." (Qualls-Corbett 1988, p. 76, 77) A positive stranger animus guides a woman into a conscious realization and appreciation of her feminine nature so she can make choices that do not compromise it. Knowing this masculine power within gives her an inner authority and she "stands constant to her feminine nature." (p. 78)

IV. The Sacred Marriage

The sacred marriage aspect of the archetype is instigated by love as an element of the divine, moving one toward a union of opposites—a realization of the Self. "You can therefore say," writes Marie-Louise von Franz, "that in every deep love experience the experience of the Self is involved, for the passion and the overwhelming factor in it come from the Self." (von Franz 1980, p. 202 quoted in Qualls-Corbett 1988, p. 83) At the intrapsychic level, when the opposites are reconciled one takes responsibility towards one's creativity and a determination "to *make* things happen, to feel new feelings, to re-image oneself." (p. 84)

In dynamic systems theory, out of personal love that "doesn't stop" there emerges the archetypal experience of Divine love. This is felt as a union with the divine, "the source and the power of love. Through the mystical union a portion of divine love is received and contained within oneself," writes Qualls-Corbett. This sacrifice to a greater authority transforms ego desires and power identification into a love beyond reason. (Qualls-Corbett 1988, p. 86) Ester Harding said a woman with this experience realizes "her body must be a worthy vessel" and she has a "power to love another [beyond that of desire]." (Harding 1971, p. 151 ff quoted in Qualls-Corbett 1988, p. 86, 87) Qualls-Corbett proposes that we may still realize the power of the divine in a psychological manner, "but only when the emotions which charge the images of the sacred prostitute, the goddess, the stranger and the sacred marriage are honored by conscious understanding." (p. 87)

V. The Four Levels of Anima Development in a Male

The anima plays a vital role in the male psyche as a personification of a male's unconscious. She can cast spells of moodiness, irritability and depression or bring joy, excitement and a sense of well being. (Qualls-Corbett 1988, p. 89) The eternal image of the woman has always mystified and inspired men through a fascination with her sexuality and spirituality. (p. 90)

A man envisions a woman as a whore if he expects sexual gratification on demand from his wife. This is the Eve, or purely biological and sexual stage, of anima development where "woman is equated with the mother and only represents something to be fertilized." (CW 16, ¶ 361)

The opposite is an elevation of the woman "to the heavenly height of The Virgin Mother; she is all things pure and holy, and consequently is untouchable." (Qualls-Corbett 1988, p. 92, 93) Jung described this stage as replacing Eve with "spiritual motherhood." (CW 16, ¶ 361) Relationship with a real woman is impossible because no woman can meet that standard, leaving the man disenchanted and disheartened. A common third face is the mother-wife, featuring the "elemental or static aspect of the feminine, associated with conservative and unchanging attitudes." The security of a stable situation may lead to a lack of the experience and emotional challenges necessary for anima development. (Qualls-Corbett 1988, p. 93) Jung writes:

> The overwhelming majority of men on the present cultural level never advance beyond the maternal significance of woman, and this is the reason why the anima seldom develops beyond the infantile, primitive level of the prostitute. Consequently prostitution is one of the main by-products of civilized marriage. (CW 10, ¶ 76)

Men are in particular danger of losing connection with the anima after mid-life, which Jung said,

> means a diminution of vitality, of flexibility, and of human kindness. The result, as a rule, is premature rigidity, crustiness, stereotypy, fanatical one-sidedness, obstinacy, pedantry, or else resignation, weariness, sloppiness, irresponsibility, and finally a childish...[lethargy] with a tendency to alcohol. (CW 9, I, ¶ 147)

Jung described a stage between Eve and the Virgin Mary as one historically represented by Helen of Troy. "[It] is still dominated by the sexual Eros, but on the aesthetic and romantic level where woman has already acquired some value as an individual." He recognized the Biblical Sophia as the fourth and highest anima state: "[It] represents a spiritualization of Helen and consequently of Eros as such. That is why *Sapientia* was regarded as a parallel to the Shulamite in the Song of Songs." (CW 16, ¶ 361) Sophia embracing her sexuality and Sophia as the highest spiritual development of the anima was described as being the bride of God. In Proverbs Sophia says, "I was daily his delight, rejoicing before him always, rejoicing in his inhabited world and delighting in the human race." (Proverbs 8: 30, 31) The Shulamite of the Bible is archetypically related to the sacred prostitute. (Qualls-Corbett 1988, p. 103, 104) (see Appendix H: The Black Goddess) To Jung she signified, "earth, nature, fertility, everything that flourishes under the damp light of the moon, and also the natural life-urge." (CW 14, ¶ 646)

The sacred prostitute is one image of a man's anima that is related to each of these four stages. Writes Qualls-Corbett, "She offers pleasure, excitement and vitality, a personification of both spirituality and earthiness. She is a lover whose beauty is exciting, whose virginal nature brings forth new life and leads to Wisdom—which is more than simply intellect." (Qualls-Corbett 1988, p.104) (The range of sexuality from pure biological eroticism through a creative spiritual dimension is also displayed by Hermes)

The affective contact with the anima is mirrored by a man's relationship with women. The experience of sex for a psychologically mature man is not that of power and control but of a feeling of honor and devotion to the feminine mystery. Qualls-Corbett cautions that the new consciousness brought about by a union with the feminine "is not achieved without frustration, conscious suffering and fear, for it is invariably accompanied by disconcerting changes." (Qualls-Corbett 1988, p. 107)

VI. The Sacred Prostitute in the Feminine Psyche

The sacred prostitute can also play a revitalizing role in the psyches of modern women. The archetype as a pattern of behavior helps to motivate, regulate and shape feminine consciousness. (Qualls-Corbett 1988, p. 118):

> The image of the goddess, associated with the physical beauty of the feminine body, activates archetypal energies pertaining to love, passion and relatedness. The woman who honors these feelings comes to understand and appreciate the sacred prostitute incarnate in her own personality. (p. 119)

If a woman doesn't appreciate and honor her body for what it is, she ends up comparing herself to the advertiser's image, thus rendering herself vulnerable to the negative animus messages like, "You'll never be good (pretty, desirable, etc.) enough!" (p. 119)

Beginning to develop a loving relation with her feminine is often expressed in women's dreams as loving or making love to a woman (Qualls-Corbett 1988, p. 127) or as a "girl child, young and fragile, whom the dreamer must nurture and protect." The positive animus, the stranger, appears at a later stage of a woman's psychological development. (p. 128)

For a woman to consciously redeem the split off or degraded aspects of her feminine nature she must begin by realizing how she is being wounded by inappropriate or repressed attitudes. These must be confronted. Then she can begin to experience the beauty and importance of her body and sexuality. This allows a sense of the goddess and the sacred prostitute to emerge which encourages the woman to love. Her ego strength grows as she develops her feminine nature, which allows her to welcome the masculine Other, her stranger animus. The animus initiates her into her essential, independent femininity, "a position of strength from which she may relate to both her outer and inner worlds. Fear of men, for guilt at using them ceases to be a problem... Such a woman simply revels in the experience of love, both the giving and receiving of it." (Qualls-Corbett 1988, p. 132)

Ester Harding writes: "The spirituality of the woman must be distilled from concrete experience; it cannot be obtained directly [like a revelation]." (Harding 1971, p. 150) Qualls-Corbett adds: "The internal stranger animus may facilitate the woman's awareness of her sexuality, but it takes an actual man to concretize the experience of love." (Qualls-Corbett 1988, p. 138)

These archetypal energies survive in the collective unconscious of a culture and can be seen in the dreams of women. (Hillel 1977, p. 69) "In a dream, in a vision of the night, when deep sleep falleth upon men, in slumbers upon the bed; Then He openeth the ears of men, and sealeth their instruction." (Job 33: 15, 16) The writings of Hillel and Qualls-Corbett present an array of dreams to illustrate the stirring of the feminine unconscious. Hillel presented one dream in particular that epitomizes healing via the erotic feminine, symbolized by the vulva: A female Jungian in the final stages of training to be an analyst dreamt that a frail and dying Jung asks to put his hand in her vagina. That act completely healed him. (Hillel 1997, p. 11)

The vulva, or vagina, was the principle symbol of the Goddess who was identified with nature over which she held sovereignty. Her dignity was located in her vulva: "The vulva was, above all else, a sacred place imbued with divine meaning. The Goddess' vulva connoted a primordial place of female power—the seat of life's affirmation, an abode of desirousness and ecstatic passions." (Hillel 1997, p. 13) Hillel sees in the sacred vulva a mysterious power where sensuality and sacredness are reconciled (p. 19) and claims, "Healing the feminine is ultimately concerned with the dignity and sacredness of the Goddess' vulva." (p. 75)

APPENDIX H

The Black Goddess

Some consider the role of the Sphinx in the myth of Oedipus to be the secret to the dis-ease in the Western psyche. Peter Redgrove in *The Black Goddess and the Unseen Real* explores the psychology, physiology and spirituality behind the symbolism of the Sphinx. He associates the realm of the unseen real with the Black Goddess with the Sphinx being one of its archetypal images. We are largely unconscious of the great extent that we participate in and are influenced by the unseen real, yet, "our unconscious mind is a living organ of perception" of that realm. (Redgrove 1987, p. xviii) The Black Goddess can be taken to its imaginal limits by applying the concept of the human as a situated robot within a dynamic systems framework to Redgrove's examination of ritual intercourse and the Whore of Babylon.

I. Oedipus and the Sphinx

The Sphinx, "with her woman's head, lion's body, serpent's tail and eagle's wings," symbolizes fate, the forces that preside over life, the dynamics of Mother Nature, and the world-goddess "with its message of death and renewal, its knowledge of invisible things, its promise of the shamanistic mastery of the two worlds of life and death, and the worlds of past and future." (Redgrove 1987, p. xxv) She was the winged moon-goddess of Thebes who reflected the moon-cycle while "her composite body was an astronomical calendar-picture of the Theban year: a lion for the waxing part and a serpent for the waning part." (p. xxiii) Egyptians saw the Sphinx with its masculine front "(with the reigning pharaoh's face) and feminine behind" as a vision of the opposites united in nature. (Massey 1883, p. 139 in Redgrove 1987, p. xxvii)

The answer to the Sphinx's riddle about the union of opposites; life-death, beast-human, etc., was either man or God, and Oedipus answered man. (Walker 1983, p. 957 in Redgrove 1987, p. xxvi) Oedipus' answer to the Sphinx caused her to "destroy her visible appearance and to cease instructing him." (p. xxiii) Jung remarked how Oedipus walked right into the Sphinx's trap by "overestimating his intellect in a typically masculine way...The riddle of the Sphinx was *herself*." (CW 5, ¶ 265) Oedipus should have contemplated the dual nature of the feminine for a man: "compounded of animal instinct (the body) and specifically human qualities (the head). In her body lie the forces that determine man's fate, in her head the power to modify them intelligently."(CW 10, ¶ 715 quoted in Redgrove 1987, p. xxvi)

Freud believed that the Oedipus complex, "or something like it, was reproduced in every Western individual." (Redgrove 1987, p. xiv) Freud's cultural bias caused him to miss "the strong female undertow" in the myth. (p. xxii) Freud, who "as a representative of our time, profoundly distrusted sexuality," said of the Oedipus blinding after realizing he fulfilled the prophesy by killing his father and sleeping with his mother: "Like Oedipus we live in ignorance of these wishes...and after their revelation we may all of us seek to close our eyes to the scenes of our childhood." (Freud 1900, p. 365 quoted in Redgrove 1987, p. xx) "Freud in effect forbids our acquiring direct knowledge from adult sexuality; he forbids 'carnal knowledge,'" says Redgrove. (p. xxix)

Redgrove suggests Thoreau and the Romantics offer an entrée to what the Sphinx represents through a sensuous relationship to our bodies and reality. Thoreau's hope was "the human being may one day 'delight in that full life of the body which it now fears.'" (Redgrove 1987, p. xxi-xxii) "Whitehead, like the Romantics,...[called] for a science 'based on an erotic sense of reality, rather than an aggressive dominating attitude towards reality,' the latter being the Oedipal atti-tude." (p. xxii) Goethe had his magician Faust call up the Earth-Spirit. He then "recoiled at its 'ugliness,' then repelled it with his arrogance: 'I am Faust, in everything thy equal!' [Goethe 1976, 1, p. 500]...Faust goes on to transformation and redemption [and] visits the mysterious Earth-Mothers who are underground." (p. xxv) The transformation occurred through the magical, Romantic solution of "continued dialogue with the powers that he had called up...even though it has led them to the darkest regions and to facing up to the hauntings created by Oedipal repression." (p. xxvi) Jung, who has been called the last of the Roman-tics (Ryley 1998, p. 140), developed active imagination as a means of dialoguing with the unknown—"the basic procedure of all magic and all art," notes Redgrove. (Redgrove 1987, p. xxvi)

II. Weather and the Seasons

The Sphinx had strong associations to cycles and the seasons but the rise of the Enlightenment in the 18th century led to the deliberate ignoring of cyclical time, "particularly in respect to the weather," says Redgrove, because it was uncontrollable. (Redgrove 1987, p. xxiv) Redgrove cites research that reveals " a profound physiological response, and therefore a finely calibrated mental and poetic response, to weather changes and to electrical patterns in wind and season...The air carries deep messages in organic form from all the growing things of the world." (p. xxix-xxx) What is ignored returns in negative form, with a modern version of a Theban plague following Oedipus' unconscious acts being the plague of undiagnosed weather caused illnesses. Repressing or thinking we can remain aloof to the body and its forces leads to being driven by them, which is Oedipus unconsciously spawning offspring with his mother. (p. xxiv)

The Romantics embraced nature as a way of rescuing the soul from the death grips of Enlightenment rationalism and the polluting and mind-numbing elements of the Industrial Revolution (see Appendix B):

> The Romantics were right to affirm that their feeling-perceptions led to an actual communion with nature and a vision of its unity, and were correct in their high estimation of the imagination, which was to Coleridge 'the organ of the supersensuous,' the vehicle of understanding of such perceptions. (Redgrove 1987, p. xxx)

III. The Animal Senses and Electromagnetic Radiation

Freud said, "The deepest root of the sexual repression which advances along with civilization is the organic defense of the new form of life achieved with man's erect gait against his earlier animal existence." (Freud 1961b, p. 43 quoted in Redgrove 1987, p. xxx) Redgrove presents animals as "one of the terms in the Sphinx's riddle." (p. 2) "The Animal

Soul is that which perceives and feels, without it he may not perceive or feel the Joys of the Universe...It is the Kingdom of Heaven on Earth." (Achad 1973, p. 87 quoted in Redgrove 1987, p. l) Animals have such remarkable sensitivities to realms invisible to humans that they rightfully have been viewed as gods, "immersed as they are in their natural continuum of hearing, touch, extended vision, EM [electromagnetic] senses, profound olfactory sensitivity and such synasesthesias as the free-floating pheromone maser." (p. 47, 48)

The central Romantic idea is that nature is a cosmic animal (*All-Tier*). (Redgrove 1987, p. 1) The Romantics spoke of a galvanic or electric force that permeated all of nature and by and through which all parts—"crystals, metals, plants and animals and in the human body"—communicated and were interconnected. (p. 1, 2) This concept and the Chinese system of *feng shui*, whose medium is called *ch'i*, is given some validity by the fact that we are surrounded by electromagnetism generated by nature. (p. 37, 38) EM is an important aspect of the unseen real that the Sphinx represents. "Everything echoes with EM," says Redgrove, "every current that flows creates an EM wave; and when an EM wave meets a shape that tunes to it, an electric current flows." (p. 36)

The radiation of particular importance to the Black Goddess is the 17 octaves of infrared spectrum that entomologist Philip Callahan calls "*natural radiation*." (Callahan 1975, p. 15-17 in Redgrove 1987, p. 40) Humans see only the highest octave of infra-red called the color red which we associate with blood and life energy. The sun and moon emit enormous amounts of IR, and lesser amounts from stars and the planet Venus. (p. 40, 41) Insects are "exemplars [of] seeing more than we see... in a radiation mode which affects us most strongly." Insect spines "see" in the infra-red and the plumes of insect antennae are really "microwave antennae very similar in structure to those devised by electronic engineers." (p. 41)

Many pheromones (external hormones) and odorous materials strongly absorb IR and UV radiation. (Redgrove 1987, p. 41, 42) Redgrove references Callahan's work in writing, "Even small electric potentials in the atmosphere will cause the molecules around almost any object to glow in the infra-red...If we could see in the IR, a field of corn in the moonlight would look like a vast array of fluorescent lights." (p. 42) Humans carry auroras about them: "Our bodies too are a living fire," Redgrove points out, "for the molecular body scents and gases that surround us are irradiated by our own warm-blooded infra-red emission" as well as by atmospheric electricity and ionization. (p. 43)

IV. Human Senses and the Black Goddess

A closer examination of many common human experiences, experiments with humans, and accounts of blind people reveal that we have vast potential abilities to become more conscious of the realm of the Black Goddess. (Redgrove 1987, chapter 2, section 1) Some people are completely successful at guessing the colors of colored paper squares by feeling the paper with their hands and using infra-red "dermo-optical perception." (p. 45) Some blind people can detect moon rise by feeling the infra-red radiation on the skin. (p. 51) The famous physiologist Bethe "could tell people apart by smell, and whether they had exercised, were emotionally excited, whether they were menstruating or ill." (p. 60) Children lose their erotic animal or non-visual senses that so engage their sense of life. During "latency" they get absorbed into the Western cultural unconscious with its Judeo-Christian mythos while their psycho-sexual development at puberty favors "an ability to form abstract visualizations... over direct expression through the body." (p. 48, 49)

There are at least 200 constituents of the "odor fingerprints" detectable by humans compared to three primary colors in vision and four major categories of taste stimuli. Measuring physiological changes show that "we are responding to these subtle influences all the time whether we know it or not." (Redgrove 1987, p. 66)

Anthropologist Edward Hall said touch is "'the most personally experienced of all sensations'" and Blake called touch the Fifth Window. Hall noted the skin's great ability "to detect and to emit radiant heat, or infra-red...[and] emotional states are reflected by changes in blood supply to different parts of the body [Hall 1969, p, 62, 55]," producing texture changes that alter infra red emission and therefore olfactory emission too. (Redgrove 1987, p. 65)

A couple embracing in a slow dance is one of the ultimate human embodiments of situated robotic concepts and dynamic systems theory. (see Appendices A-C) It begins with the unique odor of each partner reflecting their individual uniqueness—"perhaps the odour is the soul, as was once believed." Redgrove asks us to see, "with our invisible eyes, this dance between partners as a true alchemy of radiant living vessels, distilling between them their thoughts which are the same as their feelings in one formal, measured yet intimately uncontrollable act." Chemical communication between partners can completely alter and enhance perceptions. The "semiochemicals" exchanged during kissing can trigger love and desire, and the power of odors in the sweat and breath are deeply affective. "Magical ritual intercourse may depend on the conscious production of external chemical messengers by means of visualizations," Redgrove suggests. "This 'animal' thing [the power of odors] is a part of our highest nature and potential, the 'lost daughter' or animal soul who is black because we do not need eyes to see her." (Redgrove 1987, p. 67)

The power of scents and odors in relationships cannot be overstated. In Galopin's *Le Parfum de la Femme* he stated: "The purest marriage that can be contracted between a man and a woman is that engendered by olfaction and sanctioned by a common assimilation in the brain of the animated molecules due to the secretion and evaporation of two bodies in contact and sympathy." (quoted in Ellis 1942, p. 69 referenced in Redgrove 1987, p. 73) Hagen (1901, *Die Sexuelle Osphresiologie*) "emphasized the particular odours of menstruation, coition and abstinence...Daly and White in 1930 thought that human sexual odors were directly comparable to communication in insects." (Redgrove 1987, p. 70, 71) Olfactory neurons "have intimate and potent connections with emotions, behaviour and visceral function. There is a direct connection with the unconscious or semiconscious 'limbic' system in the brain." (p. 76)

Alex Comfort noted that:

> "human beings have a complete set of organs which are traditionally described as non-functional, but which, if seen in any other mammal, would be recognized as part of a pheromone system." These are the apocrine glands clustered in armpit, chest and nipple, anus, ear, eyelid, lips and public region. (Birch 1974, p. 386-396 in Redgrove 1987, p. 76, 77)

Comfort, author of *The Joy of Sex*, emphasized that "odour release... [is] enhanced at the infertile time of menstruation," and women are often randiest at this time, prompting Redgrove to question "the usual scientific assumption that sex is only for reproduction." (Redgrove 1987, p. 77)

V. Physics, Physiology, Language and Synaesthesis

Redgrove describes physical, measurable elements of extra sensuous perception, "magical," and special atmospheres. He attributes this to communication through the body via non-visual bodily feelings. These feelings arise from the bodies' sensitivities to IR, ionization in the atmosphere, pheromones and other environmental chemicals, synaesthetic (multi-media) sensual responses, and the "pheromone masers" that operates in "magical atmospheres." Such atmospheres exist where powerful rituals are being performed and in other circumstances:

> the physical enhancement of atmosphere in a church at the height of a ceremony, at the theatre or concert hall, at a football match, in a room where people have been making love, or during the mutations of a storm...In creating a ritual atmosphere, we are also creating a charged gas that is sensitive not only to ourselves, but to outside atmospheric and cosmic influences: a kind of radio set or radiation detector. (Redgrove 1987, p. 119, 120)

These physical and chemical aspects must be considered in a full accounting of an embodied human experience; the human as situated robot operating in a dynamic systems theory framework. The upper end of the sensuous embodied human experience has been described in terms associated with the divine.

Language, particularly the language of poetry, is a vital avenue for entering the God-realm, stated in DST terms as increasing the dimensionality and "symbolic density" in the psyche to such a degree that God Itself is experienced as an emergent phenomena. The essence of the poetic is based on metaphor:

> "Metaphor disorients the individual senses so that they excite and fertilize each other...In this way Poetic establishes a novel interpretation of thought and feeling... In Poetic, sight can be converted into sound and texture and even scent; single words can assume physical shape, contour, fibre; groups of words may take on meanings not implied by their grammatical relations; savour, aroma, cachet may be conveyed in texture and rhythm." (Whalley 1953, p. 155 quoted in Redgrove 1987, p. 122)

Synaesthesis can be associated with concepts of the Self as an organism where all elements are interrelated and interpenetrate each other. (see volume 1, Appendix C: The Self as Organism plus discussion of the Self in Appendix B: Bootstrapping the Archetypes in this volume) Each Greek god or goddess is the essence and center of a unique cosmos and worldview with archetypal gestalts created by temple location and construction, sacred groves, scents, foods, links to particular seasons, etc. Major appeals in the charismatic, evangelical and African-American churches are heightened emotionality and a more immediate, embodied, "God-is-touching-me-personally" sense accentuated by the sights and sounds of the participants and the music.

VI. Imagination and the Black Light

Redgrove includes Goethe's "exact sensorial fantasy" and Jung's "active imagination" as methods

> attending to the imagination in order to experience a state of unknowing which is actually charged with knowledge, to partake of which requires a profound shift of consciousness. This unknowing...is the synaesthetic plenum of the unconscious or subliminal senses, responding to a continuum of such complexity and grandeur in nature that it can only be approached by means of ignorance and emptying, with symbolic and imaginative guides, and under rules which comprehend the imagination in the Romantic sense: as a device for exploring the fringes of our knowledge, and of tuning in to hitherto unapprehended realities. Thus Goethe believed: "...every process in nature, rightly observed, wakens in us a new organ or cognition... creating in the wake of an ever-creative nature..." [Goethe quoted in Lehrs 1958, p. 84, 85] (Redgrove 1987, p. 123)

> "Who has divined the high meaning of the earthly body?" [Novalis quoted in Hartman 1966, p. 156] (Redgrove 1987, p. 115)

The Black Goddess is engaged in the psychoanalytic hour, alchemy, magic, intellectual work "or in the magical space of the poet's page or painter's canvas or in the temple, Wiccan circle, or lover's bed." (Redgrove 1987, p. 158) Training or purification is necessary to enable one to work with the unexpected, with the "gostli senses" in the black theater as "the mind empties and the vivacious images enter and the invisible musician begins to play, and be heard. And one may ask questions, and the answers appear." (p. 131, 32)

The "black light" appears in the "Divine Ipseity in Islamic mysticism: '...the light of lights (*nur al-an war*)...visible because it *brings about vision, but* [is] in itself invisible.' Its appearance heralds 'super-consciousness.'" (Corbin 1978, p. 102, 100 in Redgrove 1987, p. 120) Nuit, the Great Egyptian Goddess of the night sky, declared of herself: "'My colour [*kala*] is black to the blind, but the blue and gold are seen of the seeing.'" (Grant 1980, p. 82 in Redgrove 1987, p. 121)

The imagination Redgrove is talking about is using Hillman's "imagining heart" (see Appendix K) to approach Jung's somatic unconscious and Arnold Mindell's Dream Body. This state has ecopsychological importance since it leads "to a participation in the continuum—the hither side of which is understood as a subtle body, or a state in which one reaches out beyond the skin with something which is both within and all around." (Redgrove 1987, p. 124) Redgrove writes:

> Origen's words echo Blake's on touch: "In that spiritual body the whole of us will see, the whole hear, the whole serve as hands, the whole as feet." Synesius says that God must be approached by synaesthetic imagination, the common sense. (Redgrove 1987, p.124)

Synesius considered all the senses to be the organs of the common sense, the whole of the spirit: "the most general sensory and the first body of the soul." It proceeds from the animal, the spirit (*pneuma*) that brings imagination into play. Synesius associated "spirit" with "animal": "'by its means things divine are joined with lowest things.'" [Mead 1967, p. 86, 69-71] (Redgrove 1987, p. 124) *Pneuma*, or *Ruach*, the ancient word for breath, was "an equivalent for the spirit that arouses the imagination" and can be associated with "the power and complexity of the communications by breath and air." (p. 124)

VII. The Black Goddess in Religious Traditions

The Sphinx and her sister Black Goddesses are the outcasts throughout much of recorded history:

> [She] flies long-haired through 'Sumerian, Babylonian, Assyrian, Canaanite, Persian, Hebrew, Arabic and Teutonic mythology.' [Koltuv 1986, p. xi-xii] She was particularly shunned in Mesopotamian Semitic mythology by those

who anathematised all erotic experience except that which led to the conception of children.

Thus she became the night-demoness, the succubus-incubus, the left-hand wife-husband who consorts with those who sleep alone and who blesses or curses them with nocturnal orgasms and erotic dreams...She was the child-killing witch of the menstrual period, when the womb fills with blood instead of offspring. (Redgrove 1987, p. 117)

She is the dangerous Lilith, "Adam's first wife, who was refused equality by God, and therefore fled to the Red Sea and became a storm-demon and bred storm-demons and nightmares." (p. 118) Lilith as "shaman-istic anima" (Jung) "is that part of the Great Goddess that has been rejected and cast out in post-biblical times," says neo-Jungian analyst Barbara Black Koltuv. "She represents the qualities of the feminine self that the [heavenly] Shekhina alone does not carry." (Koltuv 1986, p. 49 quoted in Redgrove 1987, p. 167)

The Black Goddess runs like an undercurrent in the Judeo-Christian tradition, finding recognition in Solomon in Proverbs. Redgrove suggests that "'Solomon's Wisdom' meant his consort, the Queen of Sheba," was similar to Shiva's love-partner Shakti being his "Wisdom" or "energy" and perhaps

the Bible has reversed the story of the Queen of Sheba traveling to learn wisdom from Solomon. The non-biblical legends say that she was black, and came from Abyssinia, "along the road of spice"...She had a cloven foot and hairy legs, so was a kind of Sphinx, and posed difficult riddles to Solomon...[asking] "What is the water that is neither in the air nor in the river, nor in the ocean nor in the rain?" The answer to this is supposed to be "the sweat of a horse in its mane" but it is a double riddle for "women's love," or the wetness between the thighs of the Queen of Sheba, with her animal pubic mane. (Redgrove 1987, p. 69)

The Song of Solomon (Song of Songs) can be seen as being tantrically perfumed:

[It] is haunted by the Black Shulamite who has lost her lover: "I am black, but comely...While the king sitteth at his table, my spikenard sendeth forth the smell thereof. A bundle of myrrh is my well-beloved unto me: he shall lie all night betwixt my breasts." [Song of Songs 1: 5, 12] (Redgrove 1987, p. 70)

The husband considers her to be "a fountain of gardens, a well of living waters." (Song of Songs 4: 12-15)

Redgrove examines the Black Goddess beyond light in the apocryphal book, The Wisdom of Solomon. She offers "the sense of reality we desire," mediated through the senses. (Redgrove 1987, p. 70) In a goddess hymn (beginning at chapter VII), she comes to Solomon as the spirit of wisdom and *"he chose to have her rather than light"*:

> She gave Solomon an unerring knowledge of things that are: the constitution of the world, the circuits of years and the positions of the stars, the nature of living creatures and the violences of winds, the diversities of plants and virtues of roots; and she was the artificer of all the things taught. She is more mobile than any motion, and penetrateth all things, because she is a breath of the power of the Almighty, and an effulgence from everlasting light. (p. 70)

A developing awareness of our repressed sensual nature leads Redgrove to suggest the Black Goddess will be the great cultural symbol transiting us into the next age. (Redgrove 1987, p. 115) The Black Goddess in the Western tradition

> speaks in her own voice in Ecclesiasticus: "It was I who covered the earth like a mist...Alone I made a circuit of the sky and traversed the depths of the abyss...I grew like a cedar of Lebanon...I was redolent of spices...Whoever feeds on me will be hungry for more, and whoever drinks from me will thirst for more..." [Ecclesiasticus, in the version quoted by Durdin-Robertson 1975, p. 202, 203]

> There is often in her manifestations the image of a tree, and of living perfume...[reminders of] the nature and extent of our unconscious olfactory communion with the continuum of nature. (Redgrove 1987, p. 115, 116)

Trees produce a host of psychoactive pheromones and "[reduce] the normal potential [electrical] difference in humans between head and foot." (p.116)

The Holy Spirit and the Black Goddess can appear to be identical (Redgrove 1987, p. 115), with the dove emblems of the Holy Spirit related to the Indian *paravata*, the symbol of lust, and the totem of the love-goddess Aphrodite. Redgrove noted, "As *Ruach* in the Bible she hovers or broods over the waters as a creative mist. She is seen sometimes as the weather and the wind, as a storm or whirlwind...The word *Ruach* combines the meanings of 'breath, mind, spirit,' as does the Greek *pneuma*." (p. 116)

VIII. The Whore of Babylon

In Revelation the Great Whore of Babylon was called an abomination but Redgrove explains who she really was in Middle-Eastern sexual-religious practices:

> Mari-Ishatar, the Great Whore, anointed her consort Tammuz (with whom Jesus was identified) and thereby made him a Christ. This was in preparation for his descent into the underworld, from which he would return at her bidding. She, or her priestess, was called the Great Whore because this was a sexual rite of *horasis*, of whole-body orgasm that would take the consort into the visionary knowledgeable continuum. It was a rite of crossing, from which he would return transformed. (Redgrove 1987, p. 125)

In the Jesus story this was Mary Magdalene anointing him for his burial and only women being at Jesus' tomb because, as Redgrove points out, "Only women could perform these rites in the goddess's name...A chief symbol of the Magdalene in Christian art was the cruse of holy oil—the external sign of the inner baptism experienced by the Taoist." (Redgrove 1987, p. 125, 126)

Kenneth Grant, an important neo-Gnostic writer (Redgrove 1987, p. 146, 147), maintains that the cross originated as a symbol of the Mysteries of Death that were of a psychosexual nature involving "passing over into that other spirit-world of enhanced senses and returning via the 'little death' *'in full consciousness.'*" (Grant 1975 p. 39 in Redgrove 1987, p. 150):

> The knowledge the man initiated by women would return with would be *samadhi*, which was said to be shared originally by all living creatures. Instead of an Oedipus, such a man would be a re-born magician, a prophet, like Tiresias. In her article on prostitution Barbara Walker describes *samadhi* as "the unique combination of beauty and kindness called *charis* (in Latin *caritas*)...like Hindu *karuna*, a combination of mother-love, tenderness, comfort, mystical enlightenment, and sex" that was dispensed by the *devadasis* of Hindu temples and the prostitute-priestesses

in Middle-Eastern temples. "Hesiod said the sensual magic of the sacred whores or Horae 'mellowed the behaviour of men'...Communing in this way with a holy whore, man could realize the spiritual enlightenment called *hora-sis*. This word appears in the New Testament (Acts 2:17) misleadingly translated 'visions.'" [Walker 1983, p. 820, 821]...[as in] the coming of the Holy Spirit at Pentecost. (Redgrove 1987, p. 126)

Horasis links sexual with religious (or revelatory) experiences and is associated with the "'marvels of the commonplace'" and "the infinite natural wonders." (Redgrove 1987, p. 126) Plutarch noted "'a clean light together with warmth'" that remained in the soul after such sexual experiences: "'This warmth...brings to pass a marvelous and fruitful opening out...like...pores, which open to give forth persuasion and affection; but little time is needed to pass beyond the body of the loved one to pass inwards to the roots of the being, and to attach oneself to the soul, now perceptible to the cleansed vision.'" (Redgrove 1987, p. 129)

Couples have been known to experience extreme bliss in prolonged sexual play, extremely delightful experiences when touching, and even a greenish-blue-light over the skin or webs of light in the room while love-making. (Redgrove 1987, p. 127, 129) Redgrove attributes these effects to bioelectricity (p. 128) and sees it contiguous with the holy whore's *karuna* or *charis* and the whole-body *horasis* as well as and the practice of touch magic by the high priestess in witchcraft circles. (p. 127)

The Gnostics celebrated a female Christ named Charis whose menstrual blood redeemed mankind. Massey says this preceded the doctrine of purification of souls by Christ and "the Eucharist was a celebration of Charis before it was assigned to the Christ." (Massey 1900, p. 41 quoted in Redgrove 1987, p. 128) The Gnostic Marcus described the experience of Charis as the Pentecostal descent of the Holy Ghost that facilitated prophesying. (Massey 1900, p. 41 in Redgrove 1987, p. 128)

The interface between this world and the other, between the spiritual and sensual, is portrayed in Greek and Sumerian Edens where "paradises of trees [are] haunted by oracular serpents. Eden...stands for the 'littoral' or in-between zone joining this world and *that* one." (Armstrong 1969, p. 8-36 and *passim* in Redgrove 1987, p. 128) In Judeo-Christian mythology the snake in Eden is associated with evil and the Tree of Knowledge is dangerous.

IX. Love, Sex and Mysticism

The Middle English Mystics described many physical elements in their mystical experiences with synaesthesia being a common experience. There were many allegories of the sense of smell, such as a Divine Odor, and feeling touched by God was important:

> The mystic is like the bride made drunk by the lover, and this is taken from the Song of Songs...The mutual caresses of the lovers in the Song of Songs also lead to the theme of love-play between God and the soul. (Redgrove 1987, p. 130)

Experiences of the female mystics like Julian of Norwich and Margery Kempe, Redgrove states, offer "the closest connection between the Middle-English Christian mystics and visionary movements on the continent, such as the Brothers and Sisters of the Free Spirit and the Friends of God (Hieronymus Bosch was said to be a member of the former)." (Redgrove 1987, p. 129, 130)

Redgrove believes that everyone has access to these divine realms, "most particularly in the sexual embrace" when a person "comes closest to poetry and what is meant by 'the Kingdom of Heaven on Earth.'" By this means we directly experience "relaxation, vision, and enhancement of body and mind." Redgrove maintains that "because of its wholeness and power...sex for the mutual illumination of the partners" is expressly forbidden by nearly every major religion. The ancients held quite the opposite, individually and ritually practicing non-procreative love. This emerged because only humans and the higher primates experience an erotic peak not only during ovulation but again around menstruation when conception is impossible. The male experiences it "in response to the woman's cycle," notes Redgrove. The church and Darwinians only acknowledge the erotic peak at ovulation because it reproduces the species. Sex without conception is regarded as "sterile" or "whores sex," associated with demons like Lilith. (Redgrove 1987, p. 132, 133)

Poet Robert Graves said the Black Goddess "promises a new pacific bond between men and women, corresponding to a final reality of love." (Graves 1965, p. 164 quoted in Redgrove 1987, p. 134) Marie-Louise von Franz believed that none less than Thomas Aquinas became a worshiper of her in his late years. She appeared to him in a vision "in the guise of Wisdom and the bride" shortly before his death and proceeded to inspire his work. (von Franz 1966, p. 428, 159, 300, 368, 242, 192, 379 in Redgrove 1987, p. 134)

Dante was a member of a group, "the faithful followers of love," "whose antecedents were said to include the Templars" (Anderson 1980, p. 84, 85 in Redgrove 1987, p. 134):

> What is certain is that Dante's art "has its most direct origins in the discoveries of the troubadours writing in Southern France from the beginning of the twelfth century," and that for these singers..."the experience of being in love totally altered their perceptions of the world...as they were driven to explore the essence of their sexual beings, so they experienced ever higher levels of consciousness in... moments of illumination." (Anderson 1980, p. 92 quoted in Redgrove 1987, p. 134, 135)

Redgrove sees the "whole tradition of *maithuna* (Tantric visionary sexuality)" in Romance literature. It could be both sensual to the point of gross eroticism as well as being divine. Yahweh's playmate Wisdom (Sapientia) of the Solomonic books, for example, was the goddess of the Italian poets. (Redgrove 1987, p. 135)

These energies appear in Christianity in the rich tradition constructed around Mary Magdalene whom some called Mary Lucifer—"Mary the Light-Giver." The church cast Mary Magdalene as a woman repentant of her unchaste, whorish ways but other Western spiritual writings such as the *Gnostic Gospels* describe Magdalene as "Jesus' most intimate companion," often kissing her on the mouth, "and a symbol of Divine Wisdom." (Pagels 1981, p. 77 quoted in Qualls-Corbett 1988, p. 149) She was described as being "a capable, active, loving woman with the ability to know and to speak 'the All'...[with] the ability to know inexplicable things...She did not question it, as did the other disciples; she trusted her innermost source." (p. 150) Redgrove writes, "In the *Pistis Sophia*, a third-century Gnostic scripture, Mary Magdalene became the questioner of Jesus, much in the manner of the riddling of the Queen of Sheba or of the catechism of the oriental love-books between Parvati and Shiva, or between the Dark Girl and the Yellow Emperor." (Redgrove 1987, p. 135, 136) Ean Begg said Magdalene had "whore wisdom" and Redgrove associates her with "the establishment of cults of the Black Virgin, particularly in Southern France, the Cult of Love, and the Gnostic sects." Begg adds Cathars, Templars and alchemists to the list. (Begg 1985, p. 145 in Redgrove 1987, p. 136) There are over 200 famous shrines to the Black Madonna in Western Europe. The

Black Goddess is the "shadow aspect of the Madonna relating to hereti-cal knowledge." (Begg 1985, p. 198 quoted in Redgrove 1987, p. 138) "Only in India has [the cult of the Black Goddess] persisted unbroken through the centuries of Laki-Lalita, the *dark flame.*" (Begg 1984, p. 80 in Redgrove 1987, p. 138)

The Gestalt focus on being embodied, of embodied perception, and the situated robotics concepts within a DST framework reach profound dimensions through the archetype of the Black Goddess. Extra-Sensuous Perception can be generated from "nature's untranslatable concreteness" by focusing on the "non-visual bodily feelings" through "'a readiness of the body and a suspense of the will enabling total reception from the senses.'" (Hartman 1966, p. 86-89 quoted in Redgrove 1987, p. 136)

X. Dreams and the Black Light

The Black Virgin, like Isis with her temples of dream incubation, is the goddess of dreams and "all those marvels we see by inner light when our eyes are closed," earning her the occasional title of Notre-Dame du Lumiere: Black Light (Begg 1985, p. 14 in Redgrove 1987, p. 136, 137):

> She is the goddess of clairvoyance, clear-seeing and the second sight...[and] the lover's light of touch in bed...She is the Goddess of Intimacy, of being 'in touch' and of that fifth window, the skin...

> She is the Black Goddess also because she lives in the dark-ness men have created by their blindness...*She is also blind Salome, haunting Jung's autobiography.* (emphasis added, p. 137)

Working with dreams to consciously experience them is also help-ful because dreams are synaesthetic events like poetry. Dreams are like poetic language that increase

> "paradox, ambiguity, irony, tension—devices whereby the poetic imagination subverts the 'reasonableness' of language...[There is] the substantial identity between poetic logic (with its symbolism, condensation of meaning, and displacement of accent) and dream logic." [Brown 1959, p. 278, 279] However, language is capable of reaching from "depressive" to "oceanic" and back again in one statement if we will let it. (Redgrove 1987, p. 177)

> Language, which can handle all the senses, thought and
> feeling too, clears like a magic mirror to the supersensible,
> which includes one's inner senses and further personality.
> (p. 175)

Dreams help raise to consciousness the nature of unconscious influences. The powerful rhythm of the woman's menstrual cycle influences those around her and affects dream life. Therefore Redgrove notes, "It is important to relate the dream to the fertility cycle if one is a woman or living with a woman; to the moon's phases if one is not." (Redgrove 1987, p. 177)

XI. The Gnostics and Ritual Sex

Among the exercises Redgrove recommends for contacting the invisible world of the Black Goddess are a type of writing, *prynayama* breathing, and relaxation exercises. (Redgrove 1987, p. 174, 176) Many books describe exercises and rituals for magical sex as a powerful avenue to the realm of the Black Goddess. (p. 178) Certain modern gnostic sects practice magical alchemical sexuality that may be related to ancient gnosticism, the wisdom of alchemy and the Rose Cross. They celebrate a sacramental Mass of the Holy Ghost by mixing and consuming male and female essences. (p. 147, 148)

Redgrove recognizes the Great Work of all sex-alchemy orders to be "the transformation of human consciousness and its re-adjustment to a more fruitful and feminist path of evolution." (Redgrove 1987, p. 141) "Gnosis implies direct objective experience" achieved by emphasizing experience over belief and a focus on the imagination. Symbols help practitioners experience themselves and the world more deeply. (p. 142) Participants in the Gnostic communion have trained to delay or multiply their orgasms which are not equated with ejaculations. The partners remain in genital contact for extended periods of time. The "*pneuma*" generated "either was, or communicated with, their goddess." All elements of Extra-Sensuous Perception are experienced. "Poetry and prophetic speech [are] a continuum with each other and with sexual experience." (p. 142, 143)

Many Gnostics believe "the mysterious *prima materia* or starting-point for alchemy is said to be certain products of the body which in the initiate are returned and transformed by the irradiation of the 'fire snake'...These products include semen and menstrual fluid, which are reputed to be hormonally very potent, and urine...This again would be to participate like animals in the thriving world of chemical communication." (Redgrove 1987, p. 150) The Gnostic priestess is believed to see "'clairvoyantly with her womb'" during her menstrual flow within which "'the astral forces are able to assume almost tangible form'... These are the 'perfumes or *kalas* of creative breath' and the black light of the goddess bathes the priest 'in a sweet-smelling perfume of sweat.'" (Grant 1975, p. 2, 11, 12, 20 quoted in Redgrove 1987, p. 148)

The outward senses are directed inward during ritual sexual intercourse to arouse the inner world in order to "'stimulate a flow of nectar through the whole system'" whose fragrances awaken the Fire Snake, the temporary blindness when the eyes turn upward at the moment of orgasm. "'[This] occurs only when the goddess is suffused with the flow of...the nectar emitted from the flowers of the inturned sensualities, the fragrance and glow of the elixir of life itself.'" (Grant 1975, p. 70, 71 quoted in Redgrove 1987, p. 151) "'The Elixir of Life or the perfumes of regeneration'" are also part of Kali tantrism—"'a scientific experiment with the psychosexual chemistry of the human body.'" (Grant 1975, p. 73 quoted in Redgrove 1987, p. 151) The ritual sex of the Gnostics and Kali Tantric sex ["Kali is the Black Goddess"] (p. 151) would be unusually intimate and powerful transformative experiences for the embodied human imagined as the "situated robot" in a dynamic systems theory context.

Gerald Massey, an important 19th century poet and mythographer, stated that "the customs and language show that '*female influence on the sexual sense was the earliest human power acknowledged by the male,*' and maintains that he did not *worship* her any more than he did the animals, but rather recognized in her the 'embodiment of a superior potency' which was nonetheless a power he could turn to his own uses." (Redgrove 1987, p. 150) (Hermes most primal form was the phallus that was aroused by the Great Goddess):

> 'It is this radiance in women...that urges men to every sort of heroism, be it martial or poetic,' says Coomaraswamy. [Coomaraswamy 1971, p. 123]
>
> Thus Sekhet, the lioness, is not only solar heat figured to the Egyptians as a goddess but 'she is also the divinity of sexual pleasure and strong drink; the fierce inspirer of the masculine potency.' Thus Shiva had his *sakti*, or 'energy'; in

the same way, the Lord was accompanied by his 'Wisdom' before the beginning of things.

The image of Sekhet's force is her hind parts, the source of her invisible fire, and the fire-feeling; the goddess is the *energiser* of the god. "It was because the female was the *inspirer* of the breath of life, the quickener, that the spirit was considered to be a feminine nature. Even the Hebrew *Ruach* or spirit of pubescence that descended on the male at puberty is feminine in gender, as if it were the *sakti* or feminine inspirer of the male!" (Massey 1883, vol. 2, p. 272, 273 quoted in Redgrove 1987, p. 150, 151)

Excellent sex enhances the dark senses of the skin which can be shared in the temple work or "any kind of circle or healing work." Redgrove believes "nearly everybody has powers as a natural healer, and sex is a way of developing all these kinds of awarenesses that depend on pheromonal and electrical emissions from the aware body, and contact with the continuum." (Redgrove 1987, p. 179) He adds, "*The creative act and the magical act of communion with the world of reality are identical.*" (emphasis added, p. 180)

XII. Answering the Sphinx

Redgrove summarizes the process: "Thus, in our whole-body experiences, the thriving pictorial or imaginal space of our arts, our magical sex, we return and recapture for our use in *this* world lessons which we learnt in *that* one." (Redgrove 1987, p. 186) In Jungian-DST terms, if the intensity of the process is high enough, as in ritual sex, the increase in symbolic density in the archetypal realm pushes the psyche into an unstable transition state that self-organizes into an emergent Self image or other creation. There are many means of consciously engaging the unconscious like dreamwork, various forms of active imagination, etc. It's all in the hermetic journey, producing archetypal images, thoughts and behaviors along the way like hexagrams as mysterious footprints to follow on one's path through life.

APPENDIX I

Mercurius

Mercurius was the god of many of the alchemists and a name for the "spirit of matter" they sought in their opus. In late antiquity Mercurius was identical with Hermes, the Greek god of revelation, and with the gnostic Hermes Thoth. (CW 13, ¶ 239ff) Mercurius was thought of as a kind of god in the earth appearing in Jung's first dream as a phallus buried in the earth. This gnostic god was considered to be the "pneumatic Adam or spiritual man, sunk in matter, who was pictured as a *phallus*—a divine-human creative spirit hidden in the depths of matter." It is the numinosum, Marie-Louise von Franz writes, "which, for modern man, appears to have moved into the depths of the earth but in actuality lives in the depths of his own psyche." (von Franz 1975, p. 205)

Alchemists often described the spirit in matter as "'the one who sleeps in Hades' or as an imprisoned being." (von Franz 1975, p. 206 note 23) Mercurius, the god in matter, "is always a paradox containing within himself the most incompatible possible opposites," von Franz notes. He is the element quicksilver (Mercury) *and* a "philosophical" substance, "a 'dry' water or a 'divine water.' As such...taken to be the basic substance of the universe." (p. 208) He is the "light of nature, which carries the heavenly spirit within it," "a hidden 'hell-fire' in the center of the earth" *and* "the fire in which God himself burns in divine love." (CW 13, ¶¶ 256, 257 quoted in von Franz 1975, p. 208) His psychic origin was suspected by the alchemists when they defined him as "spirit" and "soul." (von Franz 1975, p. 208) He is "the spirit of the world become body within the earth" (CW 13, ¶ 261) which can be thought of as the subtle body when the spirit feels embodied, when spirit has a corporeal nature associated with it.

The element Mercury is an ideal substance upon which the unconscious can project its dual nature. Mercury is the only metal that is liquid and flowing at room temperatures. If a large drop of mercury is spilled, it will splatter into many small droplets that can be gathered together to reform one drop.

Mercury was seen *"as pneumatic [spiritual] stone [which] unites spirit and matter."* As world soul he is "that mysterious, secret something which animates and brings to life everything in the world." (von Franz 1975, p. 208) He was described as a hidden spirit of Truth or "the spirit of the Lord which fills the whole world and in the beginning swam upon the waters." (CW 13, ¶ 263) Mercurius can be deceitful, double (*duplex*) and changeable, and he "enjoys equally the company of the good and the wicked." (CW 13, ¶ 267) One is destroyed if not understanding that "above all he consists of all possible opposites." (von Franz 1975, p. 208)

Mercurius was called "the true hermaphroditic Adam," containing the four elements within himself, and as a "human-divine masculine-feminine Anthropos" he was a representation of the Self. (CW 13, ¶ 268 in von Franz 1975, p. 209) His first appearance is often as *senex* and *puer* at the same time (CW 13, ¶ 269), with the "old king" (senex) being killed off during the alchemical process and replaced by the "royal youth" (puer). (von Franz 1975, p. 209)

Mercurius was often described as the whole Trinity or as the counterpart of Christ, even as "the Logos become world" (CW 13, par 277), but sometimes identified with Lucifer and the devil. (von Franz 1975, p. 209) He was closely associated with the feminine principle because he was the son of the great mother, Nature: "The mother bore me and is herself begotten of me." (CW 13, ¶ 272) On one hand, he was particularly connected with Venus and the moon, even being described as "the most chaste virgin." (¶ 273) On the other, he was equated "with the Arcadian Hermes Kyllenios, who was worshipped as *phallus*, as *god of love and fertility* [CW 13, ¶ 278]" (von Franz 1975, p. 209), or what brings the sexes/opposites together in his alchemical appearance as Cupid with his arrow as the "dart of passion." (CW 13, ¶ 278)

The Hermes/Mercurius link with the underworld and the land of the dead was explored in the main text of this book. "[Mercurius] is at once the dark initial condition and the highest achievement, an 'earthly God' who unifies everything in himself." (CW 13, ¶ 284 in von Franz 1975, p. 209) Symbols of this all-embracing unity were a circular snake biting its tail (uroboros) or fabulous beings composed of the attributes of earth, air and water. (p. 209, 210) Jung recognized these symbolic projections of the alchemists as being an encounter with the phenomenology of an 'objective' spirit, a true matrix of psychic experience, the most appropriate symbol for which is matter. Nowhere and never has man controlled matter without closely observing its behaviors and paying heed to its laws. The same is true of that objective spirit which today we call the unconscious: it is refractory like matter, mysterious and elusive, and obeys laws which are... non-human or superhuman. (CW 13, ¶ 284)

As the son of the outer world, the macrocosm, Mercurius was extolled for his "blessed greenness" as "'a kind of germination' which 'God has breathed into created things' and from which they receive their life." (von Franz 1975, p. 211) Mercurius symbolizes the collective unconscious and all the opposites; he is a dark hidden god who is compensatory to the Christ symbol. (von Franz 1975, p. 210, 211) He becomes the chthonic counterpart to the more spiritualized image of Christ, "a fabulous being conforming to the nature of the primordial mother" via the alchemical process as symbolically described by the axiom of Maria Prophetissa (Appendix J). Jung's vision of a brightly lit Christ on the cross with a body of greenish gold was a unification of the Christ-image with Mercurius. (Jung 1961, p. 210, 211)

Mercurius also participates in the god-man motif that serves as a compensatory personification from the collective unconscious. The god-man in folklore and legends appears in such forms as Elijah in Judaism, El-Khidr in Islamic legend, and Merlin of the medieval Grail saga. Legends describe Elijah as the "incarnation of an eternal soul-substance," with the same nature as that of an angel. (von Franz 1975, p. 213) He ascends to heaven in a fiery chariot where he soars above the earth as an eagle spying out human secrets or wanders the earth in disguise and tests people. His body was said to have come from the Tree of Life, he possessed two souls at birth (Hermes duplex), and he was covered in hair like a Wildman with fiery wrappings nourished by flames. (von Franz 1975, p. 213, 214)

Elijah, like Mercury, is a symbol of the Self and is more human than Christ as a personification of the god-man type since he was begotten and born in original sin. He stimulated popular fantasy because he was one of only four human beings to attain immortality with their bodies. He is also more universal than Christ because he is associated with the personification of Allah and El-Khidr and incorporated pre-Yahwistic heathen divinities like Baal, Mithras and Mercurius. He is particularly identified with the Greek sun god Helios. El-Khidr, the "first angel of

God," replaced Elijah in Islamic legend and had many of the same traits (von Franz 1975, p. 214):

> Many popular European fairytales...begin with the formula: "In the days when our Lord still wandered the earth..." Psychologically this shows that the official god-image was felt to be too "metaphysical" and too removed to the heavenly realm. One was no longer able to meet God himself in the here and the now. The god Mercurius or Elijah or Khidr, the messenger of God, on the other hand, wanders around in the guise of a stranger who confronts the individual directly. (p. 214, 215)

APPENDIX J

The Alchemical Dictum of Maria Prophetissa

The process of developing psychological wholeness, for which Hermes is the guide, was presented numerically in an alchemical dictum attributed to the Coptess or Jewess, Maria Prophetissa: "One becomes two, two becomes three, and out of the third comes the One as the fourth." (CW 9, II, ¶ 237) The creative process of generating and manifesting the four state symbolically portrayed as the activities in the gap in Hermes' wand can be mathematically described by transition phase phenomena in complexity theory. (Appendix C).

Jung explored the saying by examining the issue of "the 3 and the 1." If four dark dots (Fig. 1) represent initial unconscious wholeness as a square (1), drawing a diagonal divides the square into two triangles (2). This symbolizes something becoming conscious, which necessitates a differentiation between two things [duality or (2) state]. One element more consciously develops while its counterpart remains in the shadowy unconscious, setting up a dynamic opposition depicted by two triangles pointing in opposite directions (3). Conscious wholeness (1 as 4) is attained by the three dots of the light triangle (3) integrating one dark dot symbolizing the opposite and unconscious triangle. (CW 9, I, ¶ 425 ff)

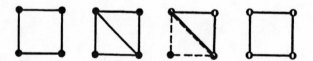

Fig. 1

The concept in the 3 state can be illustrated with psychological types where the main function is under conscious control symbolized by the white circle in the upper triangle. The two axillary functions are under varying degrees of conscious control symbolized by the half circles of white at the base of the upper triangle. The most unconscious function forms the apex of the opposing triangle, the lower triangle with the black circle symbolizing unconsciousness. Its base includes the unconscious sides of the axillary functions indicated by the dark shading of part of the circles on the base.

Another presentation of Prophetissa is to consider the (1) state to be the original wholeness of the collective unconscious, the beginning uroboros. A duality manifests (2), often through love or conflict. Pathologies and complexes become apparent as the individual or society unconsciously lurches from one extreme to the other (enantiodromia). At some point we become fully conscious, even painfully aware, of the pathological or valid positions of both sides of the conflict as they become more clearly delineated (3). Awareness is not the resolution of the conflict. When the individual or culture enters the "don't know" (Hermes') space, or deliberately seeks to attain wholeness through meditation, rituals, etc., the transcendent function may be activated. In the language of dynamic systems theory, the activity of the transcendent occurs when the psyche is in the heightened dimensionality, high energy position of the transition phase which generates new possibilities in a self-organizing manner. (see Appendices A and B) The transcendent function, operating according to its own process, timing and volition, offers a resolution in the form of a symbol that unites the opposites to form a new unity, the 4 state. This is often accompanied by a feeling of grace. A person or a culture is led to a new level of consciousness and a new worldview by using the unifying symbol as an inspiration and a guide. The symbol becomes incarnated by being lived out. It is a wholeness, but now a conscious wholeness unlike the original 1 state.

Since Hermes initially called forth the opposites from the unconscious, it is Hermes who reconnects one to the unconscious and generates the unifying symbol in the gap of his wand, illustrating Hermes association with the number 4, here the number 4 of Prophetissa. Hermes is thus the father and the son, the sennex and the puer. The ego is always challenged to differentiate *and* see relationships between things, within and without, thereby establishing order out of chaos and imparting a sense of meaning. Personal ego experiences that "don't stop" becomes the archetype of the Self as described by dynamic systems theory. (see Appendix B)

The 2 state is seen in the projection of one's faults onto others of the same sex—the shadow. After integration of the shadow (4 state of the shadow), one approaches the deeper levels of the unconscious symbolized by the archetype of the opposite sex within; the anima in males and the animus in females, what Jung called the soul archetype. This gets projected onto the opposite sex, often leading to marriage in the first half of life. With the approach of second half of life, the psyche shifts to a more spiritual orientation. Jobs and others, including one's wife or husband, are not as enchanting or engrossing as they once were: a deeper sense of meaning is desired. The soul archetype gains greater importance, and may again be projected onto a member of the opposite sex. Withdrawing the projection to its source within leads one to recognize the contra-sexual archetype as a function of the Self; to lead one toward a deep sense of wholeness (the 4 state of the anima). One's personal, individual life is challenged to develop to its fullness and be seen against the backdrop of its archetypal foundation, the Self.

The archetype symbolized by Hermes is activated again as one transits out of life. As death approaches, imagery of a union with the opposite sex can occur in dreams and visions as illustrated in the DVD, "Appointment with the Wise Old Dog" (Blum 1998) and discussed in *A Jungian Bouquet* (in preparation). The individual can now see himself or herself objectively in the context of their whole life and the broader environment as they move towards a final union with the unconscious, life's "final 4" state. Jung described this in his near death experience and von Franz commented, "Death is the last great union of the inner world-opposites, the sacred marriage of resurrection." (von Franz 1975, p. 286) Jung was critical of Freud for missing the symbolic dimensions of sex as a symbol for the union of opposites and spiritual fulfillment.

The ancient Chinese called the *conjunctio* at death the "dark union at the yellow sources." At death it was believed a person breaks up into two psychic parts: a dark Yin principle, a feminine part that sinks down to earth, and a bright Yang principle, a masculine part that ascends to heaven. The feminine part journeys to the feminine divinity of the West, the masculine part to the East to the "dark city" or to the "yellow source":

> As "Mistress of the West" and "Lord of the East" they then celebrate the "dark union" and in this union the dead man arises as a new being, "weightless and invisible," who can "soar like the sun and sail with the clouds." (von Franz 1975, p. 286)

Jung understood the dictum of Prophetissa to be a numerical presentation of the secret of alchemy. Odd numbers are archetypically associated with the "masculine, fatherly, spiritual" and even numbers with the "feminine, motherly, physical." (CW 12, ¶ 31) The alchemist's goal was to move the Christian *unity* associated with the masculine trinity (an odd number) into a feminine *unity* symbolized by the even number 4. (¶¶ 25-31) In this interpretation the masculine, over-spiritualized Christian unity (1) constellated a compensatory chthonic femininity in the collective unconscious (2) in the form of *serpens merurii*, the primordial Tiamat with her dragon attribute. (¶ 29) The alchemists engaged these contents with their opus and Mercurius emerged, symbolizing a new consciousness associated with the masculine, hence the number 3. (see Appendix I: Mercurius) This is a consciousness that retains a close connection with the feminine and when developed it produces a new unity (4) that consciously incorporates the feminine. The unity is symbolized by the sacred marriage, the *mysterium conjunctionis* of spirit and matter, masculine and feminine, and all opposites.

The Incarnation of Jesus was an incomplete attempt at 3 because it did not occur in sinful man. The 3 state appeared in the Book of Revelation when the "pagan" woman began to descend with a newborn son. Yahweh withdrew the child into the wilderness after a dragon threatened the mother and child. The Christian world would not be ready for the new level of consciousness (3 state) the child represented until the end of the Christian era, the Age of Pisces, with the dawning of the Age of Aquarius that began in the 1960s. The 4 state was experienced by many alchemists and certain enlightened individuals throughout Western history like Hildegard of Bingen, Meister Eckhart, and Jung, but in the Aquarian age the entire Western culture will move in that direction. The Aquarian symbol is that of a male water bearer pouring water into the mouth of the Southern fish, symbolizing consciousness being brought to the unconscious. (Jung 1961, p. 339, 340; von Franz 1975, p. 283, 284)

The relationship of the eternal to space-time is also symbolically presented by Maria's dictum. The original one-ness/wholeness, the original (1), is non-existence; the eternal, the Tao, the pregnant void, the *ursprung* (Gerbser). Something appears beyond, within, or out of it; a singularity occurs: a two-ness emerges (2) as yin and yang, the Mother and Father in creation myths, or Hermes as consort of the Great Mother. The interactive dance between the two generates all potential forms symbolically represented by three-ness (3). This is the creative, clarifying and shape-forming activity of the transition phase in complexity theory which symbolically occurs in the gap in Hermes' wand. Manifesting the potential forms in space-time is the four-ness (4). This is the artistic creation, final mathematical formulation, new building constructed, etc. The creation of the final form is affected by the environment it forms in and interacts with as it is forming, like the variable conditions affecting water droplets as they crystallize into unique snowflakes. Transition phase phenomena occur at each interval; from 1 to 2, from 2 to 3, and from 3 to 4.

Bringing the new whole form into existence is difficult and with a propensity to abort. There is often uncertainty and even fear of the new. There is a tendency to revert to the old familiar forms. There may be pressure from other elements in the environment to remain unchanged because these elements would also be forced to change in order to adapt to the new state of one of its members. The "difficulty of the return" is seen in Cinderella's three departures from the ball before being "fixed" in reality with the marriage to the prince. The beauty and emotional power of the new form/image can inspire one to have the courage and perseverance to manifest it in space/time.

The Dictum of Maria Prophetissa presented in sacred geometry begins with the point (1). A point has no existence; it simply delineates a spot between things. Two points (2) form the minimum line as a vector or emergence from "the Other" (1). Three points (3) make a plane and the minimal form—the triangle that exists only in the abstract as a plane between objects. The miracle of creation is going from the three to the four: four points (4) are the archetypal minimal volume—the fourth point atop the other three makes a pyramid. Four is the manifestation in space-time of an aspect or development [(2) and (3) states] of the (1).

In medieval philosophy the three-state relates to "the potential pre-existent model of creation in the mind of God, in accordance with which God later produced the creation [the four state]." The *transition* into creation occurs through the primary causes God created, either through his Son or through Wisdom, *Sapientia Dei*, "through which He knows Himself." The primary causes are *in* God and *are* God himself. Wisdom proliferates into a multitude of primordial forms ("ideas" or "prototypes") while always remaining one. (von Franz 1975, p. 247) The "prototypes" of all existing things, *Sapientia Dei* as the model of creation in the mind of God (the *archetypus mundus*), are an absolute unity where "things which are not simultaneous in time exist simultaneously outside time." (p. 247, 248)

The completion of the alchemical work has been described as the individual uniting with the *unus mundus* in the mind of God. (CW 14, ¶¶ 659 ff and ¶¶ 767 ff in von Franz 1975, p. 248) Mandala symbolism "represents a psychological analogy to the *unus mundus*" (p. 250) revealing how the multiplicity of the collective unconscious and the phenomenal world is ordered in a unity. (p. 248) The experience of the *unus mundus* was described by the alchemist Dorn as the opening of a "window on eternity," comparable to the *satori* experience of Zen Buddhism, to *samadhi* of certain Eastern teachings, or to the awakening of the Tao. An experience of the Self opens one to a transcendent realm, considerably expanding the domain of one's conscious ego state. (von Franz 1975, p. 250) Jung writes:

> If we understand and feel that here in this life we already have a link with the infinite, desires and attitudes change. *In the final analysis, we count for something only because of the essential we embody, and if we do not embody that* [the 4 state], *life is wasted.* (Jung 1961, p. 325, emphasis added)

The alchemical Mercurius is himself the *unus mundus*, "the original non-differentiated unity of the world of Being." (CW 14, ¶ 660) The concept is developed in this volume of *The Dairy Farmer's Guide* and summarized as that which happens symbolically in the gap in Hermes' wand. In dynamic systems theory this would be the heightened dimensionality in the transition phase from which emergent phenomena arise and self-organize, described mythically as Hermes calling forms into being from the Great Mother matrix of the unconscious. In accord with dynamic systems theory, the emphasis is on a process of emergence rather than constellating pre-existing potential forms. The latter would be potential forms in a quantum mechanics sense with different probabilities of distinct states emerging depending on the circumstances. In terms of the human body/brain, there are no inherited neuronal networks waiting to be activated/constellated. It may be possible to categorize the number and types of forms that emerge because of complexity theory working within the parameters of the structure and basic functioning of the human body and brain, basic needs for survival, and a similarity in the environmental influences. (Appendix C)

APPENDIX K

Archetypal Psychology and Aphrodite as the Soul of the World

James Hillman's *Archetypal Psychology: A Brief Account* (1983) presents an image-focused, mythopoetic, imaginative approach to the psyche and the world as an alternative to ego-centric, intensively subjective psychologies. Archetypal psychology modifies and extends several core Jungian concepts. It is of fundamental importance to ecopsychology because its perspective restores a sense of soul to the psyche *and to the world*. Hillman's writings on Aphrodite as the Soul of the World help us to see a world or sensuous beauty and cosmic order.

The basic premise of archetypal psychology is that the psyche expresses itself in images; images are the fundamental non-reducible elements of psyche or soul. (Hillman 1983, p. 6) "Images are the fundamentals which make the movements of psychodynamics possible." (p. 7) Psyche/soul is a self-generating activity producing dream, fantasy and poetic images. (p. 6) "Images come and go (as in dreams) at their own will, with their own rhythm, within their own fields of relations, undetermined by personal psychodynamics," Hillman writes. (p. 7) We are imagined and it is not we who imagine, making imagination "an activity of soul to which the human imagination bears witness." (p. 7, 8)

Imaginal skills must be developed to enter and work in the virtual reality of the imaginal world, as real as a dream seems to be real when dreaming it. It is a world of "fundamental structures of the imagination...[and] fundamentally imaginative phenomena." The imaginal or archetypal world "is a distinct field of imaginal realities requiring methods and perceptual faculties different from the spiritual world beyond it or the empirical world of usual sense perception and naïve formulation." (Hillman 1983, p. 3) It is a world intermediate between material/physical and spiritual/conceptual (p. 5) with its own way of knowing:

> [It teems] with transmuted substances, subtilized sensuous forms, and legions of figures each with a proper place

within the endlessly variegated topography of the *mundus imaginalis*. It is a world no longer human—or at least not exclusively or primarily human. It is *another* world, with *another* kind of reality, to which we have access through active imagination but which we explore by the exercise of an archetypal imagination...We come to know it only through the enactment of an authentically visionary imagination. (Casey 1974, p. 27)

The first presentation of an archetype is as image and its fundamental nature is accessible first to the imagination:

Its exposition must be rhetorical and poetic, its reasoning not logical, and its therapeutic aim is neither social adaptation nor personalistic individualizing but rather a work in service of restoration of the patient to imaginal realities. The aim of therapy (q.v.) is the development of a sense of soul, the middle ground of psychic realities, and the method of therapy is the cultivation of imagination. (Hillman 1983, p. 4)

The psyche's imaginal base structures the dynamics of our inner life, the forms of our perceptions, and the patterns of our relationships with the outer world, human and natural. (Hillman 1983, p. 17) Poetic fantasy is the archetypal activity of the psyche that creates our reality. (CW 6, ¶ 743) To quote Wallace Stevens, "There is always 'a poem at the heart of things.'" (Hillman 1983, p. 23) "The world of so-called hard factual reality is always also the display of a specifically shaped fantasy." (p. 23) "Fantasy is never merely mentally subjective but is always being enacted and embodied." (Hillman 1972a, p. xxxix-xl in Hillman 1983, p. 23) "The fantasy in which a problem is set tells more about the way the problem is constructed and how it can be transformed (reconstructed) than does any attempt at analyzing the problem in its own terms" (p. 45):

Imagination...is assumed to be primordially patterned into typical themes, motifs, regions, genres, syndromes... patterns [that] inform all psychic life.

If archetypal images are the fundamentals of fantasy, they are the means by which the world is imagined, and therefore they are the modes by which all knowledge, all experiences whatsoever become possible...[It] is therefore as much in the act of seeing as in the object seen, since the archetypal image appears in consciousness itself as

> the governing fantasy by means of which consciousness is possible to begin with. (p. 12)

> The fundamental ailment and loss in the West is loss of images and the imaginal sense. The result has been an intensification of subjectivity (Durand 1975), showing both in the self-enclosed egocentricity and the hyperactivism, or life-fanaticism, of Western (rather, Northern q.v.) consciousness which has lost its relation with death and the underworld. (Hillman 1983, p. 22)

Archetypal psychology starts "in the processes of imagination," not in brain physiology, language structure, societal organization or the analysis of behavior. (Hillman 1975, p. xi; Hillman 1983, p. 10) "The most fecund approach to the study of mind is thus through its highest imaginational responses [Hough 1973; Giegerich 1982; Berry 1982] where the images are most fully released and elaborated." (p. 10) Archetypal psychology is an approach through "image work" and "dream work" that *develops* soul "in its descriptions as unfathomable, multiple, prior, generative, and necessary." (p. 13) It is a poetic way of working verbally and through art, movement and play (p. 14) that reveals "imagistic universals," "that is, mythical figures that provide the poetic characteristics of human thought, feeling, and action." (p. 11) Also, "natural phenomena present faces that speak to the imagining soul." (p. 11) One must "'stick to the image' (Lopez-Pedraza) as the task at hand, rather than associate or amplify into non-imagistic symbolisms, personal opinions, and interpretations." (p. 9) This allows one to penetrate what is actually being presented and recognize the implicit involvement of a subjective stance in the process. Although an "image implies more than it presents" and has "limitless ambiguities...[which] can only be partly grasped as implications" (Berry 1974, p. 98), it presents a very specific and complex gestalt. (p. 9) "The image is a self-limiting multiple relationship of meanings, moods, historical events, qualitative details, and expressive possibilities. As its referent is imaginal, it always retains a virtuality beyond its actuality." (Corbin 1977, p. 167 in Hillman 1983, p. 9) "Image work requires both aesthetic culture and a background in myths and symbols for appreciation of the universalities of images." (p. 14)

With an archetypal approach, anything can assume archetypal significance—the stuff of our inner life and the stuff of the world. A metaphorical and imaginative response is called for that deepens and elaborates on the image. (Hillman 1983, p. 8) "All images can gain an archetypal sense" by working with an image in a way that "restore[s] psychology to its widest, richest and deepest volume so that it... resonate[s] with soul." (p. 13)

Crafting images is the equivalent of soul making, whether done concretely by an artisan or "in sophisticated elaborations of reflection, religion, relationships, social action, so long as these activities are imagined from the perspective of soul, soul as uppermost concern." Soul making is the *individuation of images*, of imaginal reality, and not individuation of the human subject. (Hillman 1983, p. 27) Examination of images and narrative details is like the explication of a text. (p. 45)

An image can be created from events and an event heard as metaphor through such manipulations as "grammatical reversals, removal of punctuation, restatement and echo, humor, amplification." This approach can be used with dreams, life situations, waking imagination or fantasy. (Hillman 1983, p. 46) "The aim of working with dreams or life events as dreams is to bring reflection to declarative and unreflected discourse, so that words no longer believe they refer to objective referents; instead, speech becomes imagistic, self-referent, descriptive of a psychic condition as its very expression (Berry 1982)." (p. 46)

Emotions and feelings are inherent in every image, as experienced in dreams. They qualify, clarify, and give force and dynamism to images while images give form and a face to the emotions. (Berry 1974, p. 63 referenced in Hillman 1983, p. 48) Images are properly approached with an awakened, imaginative heart. (Hillman 1983, p. 7) Images are animated, requiring an animal response—an immediate, reflexive response to what is being perceived in the moment. They are considered to be "neither good nor bad, true nor false, demonic nor angelic (Hillman 1977a), though an image always implicates 'a precisely qualified context, mood and scene.'" (Hillman 1977b quoted in Hillman 1983, p. 8) The effect of an image's own presentation invites a "judgment as a further precision of the image," judgments being "inherent to the image." (Hillman 1983, p. 8, 9)

"Pathologizing" and "falling apart" (Hillman 1975) are necessary for the cultural psyche to "break[ ] through self-enclosed subjectivity and restore[ ] it to its depth in soul, allowing soul to reappear again in the world of things." (Hillman 1983, p. 22) Metaphor re-animates the soul *and* the environment—"the events of the body and medicine, the ecological world, the man-made phenomena of architecture and transportation, education, food, bureaucratic language and systems." (p. 22, 23)

A crucial concept is "the human being is set within the field of soul; soul is the metaphor that includes the human" (Hillman 1983, p. 17) and "infinitely surpasses man." (Avens 1982a, p. 185 quoted in Hillman 1983, p. 17) This puts soul/psyche beyond human, personal behavior; beyond "humanistic or personalistic psychology." (p. 17) By describing soul as being primarily a metaphor, its primary rhetoric is myth, thereby locating archetypal psychology in the cultural imagination. (p. 19)

The mind has a "poetic base" and the myths are the most fundamental (archetypal) pattern of human existence. (Hillman 1983, p. 3) Myths portray the essential forms, narratives and relational patterns of the psyche (the archetypes) with the gods being exemplars of these forms. (p. 2, 3) The ego assumes a secondary position in relation to the powers and influences of the gods, requiring a reverent attitude. The gods are impersonal: relating to the gods thus removes soul-work from its subjective, personal bias. Each god and goddess represents a complete worldview and way of perceiving and responding to the world. "Perspectives are *forms* of vision, rhetoric, values, epistemology, and lived styles that perdure independently of empirical individuality," Hillman writes. (p. 34) The gods establish our value systems and basic patterns of relationship. Many Greeks did not literally believe in gods and goddesses, but realized the best and deepest expression of the *reality* of human existence is through myth and poetry about the sacred. Scientific explanations fall many dimensions short of this level of "reality." The challenge in therapy and in living in one's psyche and the world is to know what god or goddess is active at the moment: ask of an event "not *why* or *how*, but rather *what* specifically is being presented and ultimately *who*, which divine figure, is speaking in this style of consciousness, this form of presentation" (p. 34):

> In archetypal psychology, Gods are *imagined*. They are approached through psychological methods of personifying, pathologizing, and psychologizing. They are formulated ambiguously, as metaphors for modes of experience and as numinous borderline persons...Mainly, the mode of this participation is reflection: the Gods are discovered in recognizing the stance of one's perspective, one's psychological sensitivity to the configurations that dominate one's styles of thought and life. Gods for psychology do not have to be experienced in direct mystical encounter. (p. 35)

This requires a knowledge of myths and archetypal patterns, an elucidation of the mythologies and a matching, making associations, between myths, life experiences and situations in the world. (Moore 1982, Boer 1980 referenced in Hillman 1983, p. 36, 37) This brings a mythic dimension to the events of life and the experience of the natural world. Experiences become meaningful and are contained, understood, and ennobled when related to their mythic background. (Bedford 1981 in Hillman 1983, p. 20)

The gods are not transcendent, supernatural beings, but felt presences in the particulars and sensual natures in one's life and environment, inner and outer. Each god and goddess has their imperfections and pathologies, thereby providing a container, caring support, and meaning to the pathologies in our lives and "giving the dignity of archetypal significance and divine reflection to every symptom whatsoever." (Hillman 1983, p. 38) Pathologies are not considered to be evil but given, "just so" qualities of any god or goddess, of life, and of nature. A sense of depth and meaning, the realm of the gods, is usually discovered through our pathologies and not through a spiritual search, "because pathology is the most palpable manner of bearing witness to the powers beyond ego control and the insufficiency of the ego perspective." (p. 39) Lead and the stone that the builders rejected (alchemy), "blessed are they that mourn" (Christianity)—pathologies destroy the ego's sense of control and dominance, emptying it of its self-assumed importance.

The ego is not to be thought of as a unity but as a multiple (CW 8, ¶ 388 ff), an inhabitant of a polytheistic imaginal world of many gods. Hillman sees polytheism as providing "the most accurate model of human existence" because it explicates its innate diversity and "provide[s] fundamental structures and values for this diversity." (Hillman 1983, p. 32, 33) Jung wrote in his *Seven Sermons to the Dead*, "The multiplicity of the gods correspondeth to the multiplicity of man.... Man shareth in the nature of the gods." (Jung 1961, p. 386) Edward Casey put Jung's statement into an ecological and complexity theory context when he said there are endless variations of themes, but "gods or archetypes may be numerable *in particular groupings*, e.g., in given mythical situations" and "are...determinate and meaningful in relation to *other* imaginal positions" (Casey 1974, p. 13):

> Each such system will contain a finite (but not necessarily specified) number of members, each of which derives its

symbolic meaning from two factors: (1) its own intrinsic, auto-iconic (i.e., self-resembling, non-repeatable) nuclear signification; (2) its relationship with the other members of the mini-system in question (which is how its locus in imaginal space is determined). (p. 13, 14)

The multiple nature of consciousness is illustrated by the ego's position in a dream, if the ego is even present: it represents one position; one stance and perspective, among an ecology of other positions. The elements of a dream are not to be seen as parts of our personality, but distinct entities and characters in their own right (CW 14, ¶ 753):

"Personifying or imagining things" (Hillman 1975, p. 1-51)...encourages animistic engagement with the world... [and] allows the multiplicity of psychic phenomena to be experienced as voices, faces, and names. Psychic phenomena can then be perceived with precision and particularity, rather than generalized in the manner of faculty psychology as feelings, ideas, sensations, and the like. (Hillman 1983, p. 52)

Consciousness is given with "partial personalities," not to be regarded as "split off fragments of the 'I'." (Hillman 1983, p. 52) "In other cultures these multiple personalities have names, locations, energies, functions, voices, angel and animal forms, and even theoretical formulations as different kinds of soul." (p. 51, 52) Their consciousness "is demonstrated by their interventions in ego control, i.e., the psychopathology of everyday life (Freud), disturbances of attention in the association experiments (Jung), the willfulness and aims of figures in dreams, the obsessive moods and compulsive thoughts that may intrude during any *abasssment du niveau mental* (Janet)" (p. 52):

Since 'archetypal' connotes both intentional force (Jung's "instinct") and the mythical field of personifications (Hillman's "Gods"), an archetypal image is animated like an animal (one of Hillman's frequent metaphors for images) and like a person whom one loves, fears, delights in, is inhibited by, and so forth. As intentional force and person, such an image presents a claim—moral, erotic, intellectual, aesthetic—and demands a response. It is an "affecting presence" (Armstrong 1971) offering an affective relationship. It seems to bear prior knowledge (coded information) and an instinctive direction for a destiny, as if prophetic, prognostic. Images in "dreams mean well for us, back us up and urge us on, understand us more deeply than we understand ourselves, expand our sensuousness and spirit, continually make up new things to give us—and this feeling of being loved by the images."(Hillman 1979, p. 196) (p. 13, 14)

The dream as the paradigm of the psyche represents a psyche that is

fundamentally concerned with its imaginings and only secondarily concerned with subjective experiences in the dayworld which the dream transforms into images, i.e., into soul. The dream is thus making soul each night. Images become the means of translating life-events into soul, and this work, aided by the conscious elaboration of imagination, builds an imaginal vessel, or "ship of death" (a phrase taken from D. H. Lawrence), that is similar to the subtle body"...[Soul-as-dream] shows its inattention to and disregard for mortal experience as such, even for physical death itself, receiving into its purview only those faces and events from the mortal world that bear upon the opus of its destiny. (Hillman 1983, p. 28)

Hillman offers the soul as the primary metaphor for psychology and "it is psychology's job...to provide soul with an adequate account of itself." "A second task of psychology is to *hear psyche speaking through all things of the world, thereby recovering the world as a place of soul* (q.v. soul-making)."(emphasis added) (Hillman 1983, p. 16) The soul is imagined to have "polyform time"—"to be discontinuous," and to have "many avatars, many kinds and modes." (p. 43) Archetypal psychology is about soul-making (p. 4), the soul being a metaphor, a perspective and viewpoint on the events in one's life and environment. (Hillman 1975, p. x) Soul-as-metaphor

performs as does a metaphor, transposing meaning and releasing interior, buried significance. Whatever is heard with the ear of soul reverberates with under- and overtones (Moore 1978)...The metaphorical mode of soul is "elusive, allusive, illusive" (Romanyshyn 1977), undermining the very definition of consciousness as intentionality and its history as development. (Hillman 1983, p. 21)

It is important to distinguish between soul and spirit. If imagining is the natural activity of the soul, its "native dominants of fantasy structures" (the gods) condition our subjective perspectives. The only objectivity possible (itself a perspective) is to regard our own regard, "examining [our] own perspective for the archetypal subjects (q.v.

personifying) who are at this moment governing our way of being in the world among phenomena." (Hillman 1983, p. 24) Psychology cannot be an objective science because "objectivity is itself a poetic genre (similar to 'writer-as-mirror' in French naturalism), a mode that constructs the world so that things appear as sheer things (not faces, not animated, not with interiority), subject to will, separate from each other, mute, without sense or passion." (p. 24, 25) The stance of spirit appears as scientific objectivity, metaphysics, theology and philosophy. (p. 25) It includes "[the Saturnian] rhetoric of order, number, knowledge, permanency and self-defensive logic"; Apollonic rhetoric of clarity and detached observation; a "monotheistic" stance of unity, ultimacy, and identity; and the heroic and "puer" stances:

> While recognizing that the spirit perspective must place itself above (as the soul places itself as inferior) and speak in transcendent, ultimate, and pure terms, archetypal psychology conceives its task to be one of imagining the spirit language of "truth," "faith," "law," and the like as a rhetoric of spirit, even if spirit is obliged by this same rhetoric to take its stance truthfully and faithfully, i.e., literally. (Hillman 1983, p. 25)

Archetypal psychology realizes we "can never be purely phenomenal or truly objective" (Hillman 1983, p. 24) and it "eschew[s] borrowings from meditative techniques and/or operant conditioning, both of which conceptualize psychic events in spiritual terms." (p. 25, 26)

A sense of place, so important in ecopsychology, has an imaginal dimension. Casey (1982) maintained that "place is prior to the possibility of thought—all thought must be placed in order to be." Freud's "Vienna" and the "California schools" place ideas in a geographical image—fantasy locations and not merely sociological and historical contexts. (Hillman 1983, p. 30) The main psychologies originated in Northern Europe as "a necessity of a post-reformational culture that had been deprived of its poetic base" (p. 30) whereas archetypal psychology is "placed" in Southern Europe. "South" includes the Mediterranean Gods and Goddesses, "its sensual and concrete humanity...its tragic and picaresque genres (rather than the epic heroism of the North); and it is a symbolic stance 'below the border' which does not view that region of the soul only from a northern moralistic perspective." The location of the unconscious "up north" appears "as Aryan, Apollonic, Germanic, positivistic, voluntaristic, rationalistic, Cartesian, protestant, scientistic, personalistic, monotheistic, etc." (p. 31) The Southern cultural history re-orients consciousness toward non-ego factors in what are usually considered to be "Eastern'"positions:

> the multiple personifications of the soul, the elaboration of the imaginal ground of myths, the direct immediacy of sense experience coupled with the ambiguity of its interpretation, and the radically relative phenomenality of the 'ego' itself as but one fantasy of the psyche...[Avens (1980, 1982a, b) showed that] archetypal psychology is nothing less than a parallel formulation of certain Eastern philosophies. Like them, it too dissolves ego, ontology, substantiality, literalisms of self and divisions between it and things—the entire conceptual apparatus which northern psychology constructs from the heroic ego and in its defense—into the psychic reality of imagination experienced in immediacy. The 'emptying out' of Western positivisms, comparable to a Zen exercise or a way of Nirvana, is precisely what archetypal psychology has effectuated. (p. 31, 32)

"Irony, humor, and compassion" are aspects of a mature personality

> since these traits bespeak an awareness of the multiplicity of meanings and fates and the multiplicity of intentions embodied by any subject at any moment. The 'healthy personality' is imagined less upon a model of natural, primitive, or ancient man with its nostalgia, or upon social-political man with its mission, or bourgeois rational man with its moralism, but instead against the background of artistic man for whom imagining is a style of living and whose reactions are reflexive, animal, immediate. (Hillman 1983, p. 53)

It is a model that stresses certain values of personality: "sophistication, complexity, and impersonal profundity, an animal flow with life disregarding concepts of will, choice, and decision; morality as dedication to crafting the soul (soul-making, q.v.); sensitivity to traditional continuities; the significance of pathologizing and living at the 'borders'; aesthetic responsiveness." (Hillman 1983, p. 53)

Archetypal psychology offers important new perspectives for ecopsychology. The ecopsychological significance of image-focused psychology becomes apparent when the approach is "extended into the sensate world of perceptual objects and habitual forms—buildings, bureaucratic systems, conventional language, transportation, urban environment, food, education." (Hillman 1983, p. 46) This is a bold attempt to recuperate "the *anima mundi* or soul of the world by scrutinizing the face of the world as aesthetic physiognomy." (p. 47) Archetypal psychology sees the world as "the vale of Soul-making" (John Keats): "it does not seek a way out of or beyond the world toward redemption or mystical transcendence, because 'The way through the world is more difficult to find than the way beyond it' (Wallace Stevens, "Reply to Papini")." (p. 26) "The soul in the world...is also the soul of the world (*anima mundi*)" and soul-making is engaged "by taking any world event as also a place of soul." (p. 26) The fantasy of depth in archetypal psychology is that all things have an interiority that manifests by the physiognomic character they present. Psychological penetration and a reading of each event for something deeper, something beyond the "merely evident and natural," is a "psychological process that reveals its archetypal significance and interiority." (p. 29) Hillman writes, "The fantasy of hidden depths ensouls the world and fosters imagining ever deeper into things." (p. 30)

The ecopsychological aspect of archetypal psychology is particularly well illustrated by Hillman's move to restore the classical Greek and neo-Platonic concept of Aphrodite as the Soul of the World. Aphrodite, the Greek goddess of love and sensual beauty, was imported into Greek mythology in a wholesale manner from Near Eastern Goddess worship. (see Appendix G: the Sacred Prostitute and the Erotic Feminine) A goal in the Greek healing temples, the Aesclepions, was to know what god or goddess to follow in order to be healed. Hillman suggests we as individuals and as a culture can return a sense of the sacred and the soul to the world, can re-soul the world, by trying to see the world as a follower of Aphrodite. The world then becomes sensuous, beautiful and alive. It shifts from lifeless "things" to animate subjects, soul-full presences that approach from without, liberating soul from the sense of being a totally private, inner experience. The appropriate response and a way of generating this perspective is to perceive and react to the world as lover to its beloved, a response with the detailed, poetic observations of a lover and not the statistically cold, removed, analytic eye of the scientist. (Hillman 1992, p. 89-130)

Changing our perspective, re-ensouling the world, can begin by seeing the world through Aphrodite's eyes, worshiping at her temple, looking out at the world from the heart of that temple. Perspective is all important. As Hillman warned, "Ecology movements, futurism, feminism, urbanism, protest and disarmament, personal individuation cannot alone save the world from catastrophe inherent in our very idea of the world." (Hillman 1992, p. 126)

In Aphrodite's worldview we are not alone in the universe, but alive in a cosmos. The root word of cosmos is related to cosmetics, that which women use to adorn themselves. Cosmos is the beauty of the display of things and the proper arrangement and order that emerges from their interactions. This implies aesthetic as well as moral values. (Hillman 1989, p. 21) Each thing communicates its true nature through its unique inherited form and manner of sensuous display, offering particular types of relationship to other subjects. (p. 28) A tree, for example, can offer a nesting site for certain bird species, food for particular species of insects, and a jungle gym for kids. It is up to us to observe carefully and with feeling and thereby align ourselves closely enough to sense the depth of "the other." This requires an imaginative metaphoric dimension in our sensing, thinking and feeling, what Hillman calls "the thought of the heart." (Hillman 1992) Stones and plants, for example, have much to share with people: they are related to us and we to them in our own way. Out of an I-Thou relationship may come a constellation of, or an emergence of (dynamic systems theory), stone and plant natures within us as the object becomes a fascinating and mysterious "other."

Compare Aphrodite's cosmos to our concept of universe. Universe is associated with oneness with a bias toward universals, general principles and abstractions. Discrete entities are defined by their composition and usefulness, how they fit into the big picture, their ancestry and evolution over time:

> From the perspective of universe an orderly arrangement depends on the logic of whole and part and their external relations. Phenomena are imagined within the universe as a whole. Molecules, planets, forces, all events whatsoever become merely partial and so must be knit together by external relations to form a unified field theory or coherent universe. Even the transcendent God of most theologies connects with the world's variegated events by means of external relations, as creator of phenomena, their sustainer or their redeemer...Events can only be partial, insufficient, dispensable, unnecessary...They cry out to be fitted in, finding salvation from their contingency and their externality only within the embracing whole, "the universe," which alone can give them their meaning, their beauty, truth and moral value. Phenomena as such need explanations; they

do not provide their own reason, their own inherent intelligibility... A sufficient account would require infinite space and time so that the explanation of the universe would itself fall into the traps set by the notion of the universe. (Hillman 1989, p. 22)

This parallels spiritual generalizations like emanations from an oversoul or grounding in an immanent soul, or abstract truth, symbolic logic and concepts like pure Being or a hidden God. (p. 25, 29)

Cosmos, by contrast, is concerned with sensate particulars, an eachness of things. The responsiveness of the particular elements to the sensuous beauty of the form, display and behavior of each other in a given environment generates a unique quality and order to the events and space at any moment in time. With Aphrodite as the Soul of the World, "Soul is precisely the eachness of everywhere at any instant in any thing in its display as a phenomenon. *And only in this eachness does soul exist and cosmos show.*" (Hillman 1989, p. 30) "Cosmos becomes the interiority things bring with them rather than the empty universal envelope into which they must be brought." (p. 31) Soul is intrinsic in the idea of cosmos, not a great emanation or immortal nature, not some uber-Soul, transcendent element or Ground of Being, but soul is what emerges in the moment in the scrupulous, animal-like attention to detail to the elements in the environment. This is consistent with the ideas of archetypes as emergent phenomena as presented in Appendix B and cosmos as an expression of an organismic concept of the Self as presented in volume 1, Appendix A of *The Dairy Farmer's Guide*.

In this worldview suffering is "a necessity inherent in, and fitting with, cosmic order"—it has no inherent purpose. Every event has a shadow; decay is an aspect of cosmic order as Buddhists see it. "Each thing, including Blake's grain of sand, can hurt and be hurt, for each thing, to be true, good and beautiful must also be pathological." (Hillman 1989, p. 32)

REFERENCES

Abraham, F. 1995. "Chaos, courage, choice, and creativity." *Reflections*. p. 65-70.

Alkon, D. 1989. "Memory storage and neural systems." *Scientific American* 261: 42-50.

Armstrong, R. 1971. *The Affecting Presence*. University of Illinois Press: Urbana.

Avens, R. 1980. *Imagination Is Reality: Western Nirvana in Jung, Hillman, Barfield and Cassierer*. Spring Publications: Dallas.

—— 1982a. "Heidegger and archetypal psychology." *International Philosophical Quarterly* 22: 183-202.

—— 1982b. *Imaginal Body: Para-Jungian Reflections on Soul, Imagination and Death*. University Press of America: Washington, D. C.

Barton, S. 1994. "Chaos, self-organization, and psychology." *American Journal of Psychology* 49: 5-14.

Bedford, G. 1981. "Notes on mythological psychology." *Journal of the American Academy of Religion* 49: 231-247.

Berry, P. 1974. "An approach to the dream." *Spring* 1973: 58-79.

—— 1982. *Echo's Subtle Body*. Spring Publications: Dallas.

Blum, David. *Appointment with the Wise Old Dog*. 1998/2010. David Blum, star and director. Produced by Sarah Blum. http://www.davidblummusicendanddreamer.com/

Bly, Robert. 1990. *Iron John: A Book About Men*. Addison-Wesley Publishing Co.: NY.

Boer, C., trans. 1970. *The Homeric Hymns*. 2nd ed. Spring Publications: Dallas.

—— 1980. *Marsilio Ficino: The Book of Life*. Spring Publications: Dallas.

Bolen, J. 1989. *Gods in Everyman: Archetypes that Shape Men's Lives*. Harper & Row: NY.

Brown, N. O. 1959. *Life Against Death: the Psychoanalytic Meaning of History*. Sphere Books: London 1968.

—— 1969. *Hermes the Thief: The Evolution of a Myth*. Vintage Books: NY.

Bullfinch, T. 1962. *The Age of Fable*. New American Library.

Casey, E. 1974. "Toward an archetypal imagination." *Spring* 1974: 1-32.

—— 1982. "Getting placed: soul in space." *Spring* 1982: 1-25.

Chevalier, J. and Gheerbrant, A. 1994. *A Dictionary of Symbols*. John Buchanan-Brown, trans. Blackwell Publishers: Oxford.

Chodorow, J. 1991. *Dance Therapy and Depth Psychology: The Moving Imagination*. Routledge: London and New York.

Chomsky, N. 1965. *Aspects of the Theory of Syntax*. The MIT Press: Cambridge.

Corbin, H. 1977. *Spiritual Body and Celestial Earth*. Bollingen Series. Princeton University Press: Princeton.

Crutchfield, J.P., Farmer, J.D., Packard, N.H., and Shaw, R.S. 1986. "Chaos." *Scientific American* 255: 46-47.

Darwin, C. 1872. *The Expression of the Emotions in Man and Animals*. University of Chicago Press: Chicago and London. 1965, fifth impression 1974.

Deacon, T. W. 1997. *The Symbolic Species: The Co-evolution of Language and the Brain*. W. W. Norton: NY.

—— 2003. Multilevel Selection in a Complex Adaptive System: The Problem of Language Origin. In R. H. Weber and D. J. Depew, eds., *Evolution and Learning: The Baldwin Effect Reconsidered*. The MIT Press: Cambridge MA, p. 81-106.

De Vries, A. 1974. *Dictionary of Symbols and Imagery*. North–Holland Publishing: Amsterdam and London.

Durand, G. 1975. *Science de l'homme et Tradition*. Berg International: Paris.

Fay, C. 1977. *Movement and Fantasy: A Dance Therapy Model Based on the Psychology of C. G. Jung*. Master's Thesis, Goddard College, Vermont.

Freeman, W. 1990. Non-linear Neural Dynamics in Olfaction as a Model for Cognition. In E. Basar, ed., *Chaos in Brain Function*. Springer-Verlag: Berlin.

—— 1991. "The physiology of perception." *Scientific American* 264: 78-85.

—— 1994. "Neural networks and chaos." *Journal of Theoretical Biology* 171: 13-18.

Gardner, M. 1974. "Mathematical games: the combinatorial basis of the 'I Ching,' the Chinese book of divination and wisdom." *Scientific American* January 1974, p. 108-113.

Giegerich, W. 1982. Busse fur Philemon: Vertiefung in das verdorbene Gast-Spiel der Gotter. In *Eranos Jahrbuch* 51.

Gleick, J. 1987. *Chaos: Making a New Science*. Penguin Books: NY.

Goldberg, B. 1930. *The Sacred Fire: The Story of Sex in Religion.* Horace Liveright: NY.

Grandin, T. 2005. *Animals in Translation.* Simon & Schuster: NY.

Grandin, T. and Barron, S. 2005. *Unwritten Rules of Social Relationships.* Future Horizons: Arlington, TX.

Graves, R. 1960. *The Greek Myths.* Vol. 1. Penguin Books: New York and Harmondsworth, England.

Hamilton, E. 1942. *Mythology.* Little, Brown & Co.: Boston.

Harding, M. 1971. *Woman's Mysteries: Ancient and Modern.* Harper Colophon Books: NY.

Hastings, J., ed. 1956. *Encyclopedia of Religion and Ethics.* Vol. 6. T. and T. Clark: Edinburgh.

Haule, J. 2010. *Divine Madness: Archetypes of Romantic Love.* Fisher King Press, Carmel.

Henderson, J. 1984. *Cultural Attitudes in Psychological Perspective.* Inner City Books: Toronto.

Hendriks-Jansen, H. 1996. *Catching Ourselves in the Act: Situated Activity, Interactive Emergence, Evolution, and Human Thought.* MIT Press: Cambridge.

Hillel, R. 1997. *The Redemption of the Feminine Erotic Soul.* Nicolas-Hays: York Beach, Maine.

Hillman, James. 1972. An Essay on Pan. In *Pan and the Nightmare* (with W. H. Roscher). Spring Publications. p. i-lxiii.

—— 1975. *Re-Visioning Psychology.* Harper & Row: NY.

—— 1977a. "The pandaemonium of images: C. G. Jung's contribution to Know Thyself." *New Lugano Review* 3: 34-45.

—— 1977b. "An inquiry into image." *Spring* 1977: 62-88.

—— 1979. *The Dream and the Underworld.* Harper & Row: NY.

—— 1983. *Archetypal Psychology: A Brief Account.* Spring Publications: Dallas.

—— 1989. "Cosmology for soul: from universe to cosmos." *Sphinx* 2: 17-33.

—— 1992. *The Thought of the Heart and the Soul of the World.* Spring Publications: Woodstock, Conn.

Hogenson, G. 2000. "Archetypes as emergent phenomena." Paper delivered to the Chicago Society of Jungian Analysts.

—— 2001. "The Baldwin Effect: a neglected influence on C. G. Jung's evolutionary thinking." *Journal of Analytical Psychology* 46: 591-611.

—— 2003. "Reply to Maloney." *Journal of Analytical Psychology* 48: 107-116.

—— 2004a. "The self, the symbolic and synchronicity: virtual realities and the emergence of the psyche." Paper presented to the Chicago Society of Jungian Analysts.

—— 2004b. "What are symbols symbols of? Situated action, mythological bootstrapping and the emergence of the self." *Journal of Analytical Psychology* 49 (1): 67-81.

—— 2005. "The self, the symbolic and synchronicity: virtual realities and the emergence of the psyche." *Journal of Analytical Psychology* 50: 271-285.

Hogenson, G., Stevens, A., and Ramos, D. 2003. "Debate: psychology and biology." *Cambridge 2001: Proceedings of the Fifteenth International Congress for Analytical Psychology.* Daimon Verlag: Einsiedeln, Switzerland. p. 367-377.

Homer. 1966. *The Odyssey*. E.V. Rieu, trans. Penguin: Harmondsworth.

Hooke, S. 1963. *Babylonian and Assyrian Religion*. University of Oklahoma Press: Norman.

Hough, G. 1973. "Poetry and the anima." *Spring* 1973: 85-96.

Hubel, D., Wiesel, T., and LeVay, S. 1977. "Plasticity of ocular dominance columns in monkey striate cortex." *Philosophical Transactions of the Royal Society of London* 278: 377-409.

Jung, C. *The Collected Works of C. G. Jung.* [CW] 2nd ed. H. Read, M. Fordham, G. Adler and W. McGuire, eds., R.F.C. Hull, trans. Princeton University Press: Princeton, NJ.

—— CW 5. 1956. *Symbols of Transformation.*

—— CW 6. 1971. *Psychological Types.* H. G. Baynes, trans., revised by R. F. C. Hull.

—— CW 8. 1969. *The Structure and Dynamics of the Psyche.*

—— CW 9, I. 1968. *The Archetypes and the Collective Unconscious.*

—— CW 9, II. 1968. *Aion.*

—— CW 10. 1970. *Civilization in Transition.*

—— CW 12. 1968. *Psychology and Alchemy.*

—— CW 13. 1970. *Alchemical Studies.*

—— CW 14. 1970. *Mysterium Coniunctionis.*

—— CW 16. 1982. *The Practice of Psychotherapy.* 1st ed. H. Read, M. Fordham, and G. Adler, eds., R.F.C. Hull, trans.

—— CW 18. 1976. *The Symbolic Life: Miscellaneous Writings.*

—— 1961. *Memories, Dreams, Reflections.* Aniela Jaffe, ed., Richard and Claire Winston, trans. Random House: New York.

—— Ed. 1964. *Man and His Symbols.* Doubleday and Co.: Garden City, NY.

—— 1977. *C. G. Jung Speaking: Interviews and Encounters.* William McGuire and R.F.C. Hull, eds. Princeton University Press: Princeton, NJ.

Kandel, E. 1989. "Genes, nerve cells, and the remembrance of things past." *Journal of Neuropsychiatry and Clinical Neuroscience* 1: 103-125.

Kaufman, S. 1993. *The Origins of Order.* Oxford University Press: NY.

Kerenyi, K. 1976. *Hermes Guide of Souls.* Murray Stein, trans. Spring Publications: Zurich.

King C. 1991. "Fractal and chaotic dynamics in nervous systems." *Progress in Neurobiology* 36: 279-308.

Lauch, J. 2002. *The Voice of the Infinite in the Small: Re-Visioning the Insect-Human Connection.* Shambala: Boston and London.

Lewin, R. 1992. *Complexity: Life at the Edge of Chaos.* Macmillan: NY.

Lopez-Pedraza, R. 1977. *Hermes and his Children.* Spring Publications: Zurich.

Lutzenberger, W., Elbert, T., Birbaumer, N., Ray, W., and Schupp, H. 1992. "The scalp distribution of the fractal dimension of the EEG and its variation with mental tasks." *Brain Topography* 5: 27-34.

Merritt, D. L. 1988. Jungian Psychology and Science: A Strained Relationship. In *The Analytic Life.* Sigo Press: Boston.

Milton, J.G., Longtin, A., Beuter, A., Mackey, M.C., and Glass, L. 1989. "Complex dynamics and bifurcations in neurology." *Journal of Theoretical Biology* 138: 129-147.

Mithen, S. 1996. *The Prehistory of the Mind: The Cognitive Origins of Art, Religion and Science.* Thames and Hudson: London.

Moore, T. 1982. *The Planets Within.* Bucknell University Press: Lewisburg, Pennsylvania.

Mpitsos, G. 1989. Chaos in Brain Function and the Problem of Nonstationarity: a Commentary. In E. Basar, T.H. Bullock, eds. *Brain Dynamics.* Springer-Verlag: Berlin.

Neumann, E. 1955. *The Great Mother: An Analysis of the Archetype.* Ralph Manheim, trans. Pantheon Books: NY.

Nicholas, J. 1986. Chaotic Dynamics in Biological Information Processing: A Heuristic Outline. In H. Degn, A.V. Holden, L.F. Olsen, eds. *Chaos in Biological Systems*. Plenum Press: NY.

Otto, W. 1981. *Dionysus: Myth and Cult*. Spring Publications: Dallas.

Paris, G. 1990. *Pagan Grace: Dionysos, Hermes, and Goddess Memory in Daily Life*. Spring Publications: Dallas.

Pfeiffer, R. and Scheier, C. 1999. *Understanding Intelligence*. The MIT Press: Cambridge, MA.

Piaget, J. 1962. *Play, Dreams and Imitation in Childhood*. W. W. Norton & Co.: NY.

Plaut, A. 1966. "Reflections about not being able to imagine." *Journal of Analytical Psychology* 11.

Qualls-Corbett, N. 1988. *The Sacred Prostitute: Eternal Aspect of the Feminine*. Inner City Books: Toronto.

Redgrove, P. 1987. *The Black Goddess and the Unseen Real*. Grove Press: NY.

Rilke, R. 1934. *Letters to a Young Poet*. W. Norton: NY.

Romanyshyn, R. 1977. "Remarks on the metaphorical basis of psychological life." Paper in the First International Seminar on Archetypal Psychology, University of Dallas.

Scarf, M. 1986. "Intimate partners: patterns in love and marriage." *The Atlantic Monthly*, Dec., 1986, p. 66-76.

Schmidt, G. 1995. "The chaotic brain: a unifying theory for psychiatry." Unpublished paper delivered to the Department of Psychiatry, Univ. of Wisconsin, Madison, WI.

Schwartz-Salant, N. 1982. *Narcissism and Character Transformation*. Inner City Books: Toronto.

Sejnowski, T., Koch, C., and Churchland, P. 1988. "Computational neuroscience." *Science* 241: 1299-1306.

Sheldrake. R. 1999. *Dogs That Know When Their Owners are Coming Home*. Three Rivers Press: NY.

Skarda, C., and Freeman, W. 1987. "How brains make chaos in order to make sense of the world." *Behavioral and Brain Sciences* 10: 161-195.

Stassinopoulos, A. and Beny, R. 1983. *The Gods of Greece*. Harry N. Abrams: NY.

Stewart, L. H. 1977. "Sand Play Therapy: Jungian Technique." *International Encyclopedia of Psychiatry, Psychology, Psychoanalysis and Neurology*. B. Wolman, ed. Aesculapius Publishers: NY. p. 9-11.

—— 1978. Gaston Bachelard and the Poetics of Reverie. In *The Shaman from Elko*. G. Hill et al, eds. C. G. Jung Institute of San Francisco: San Francisco.

—— 1981a. Play and Sandplay. In *Sandplay Studies: Origins, Theory and Practice*. G. Hill, ed. C. G. Jung Institute of San Francisco: San Francisco. p. 21-37.

—— 1981b. "The play-dream continuum and the categories of the imagination." Presented at the 7th annual conference of the Association for the Anthropological Study of Play. Forth Worth: April 1981.

—— 1982. Sandplay and Analysis. In *Jungian Analysis*. M. Stein, ed. Open Court Publishing Co.: La Salle, IL. p. 204-218.

—— 1984. "Play-eros, in affects and archetypes II." Paper presented at active imagination seminar in Geneva, Switzerland in August 1984.

—— 1985. "Affect and archetype: a contribution to a comprehensive theory of the structure of the psyche." In *The Proceedings of the 1985 California Spring Conference*. C. G. Jung Institute: San Francisco. p. 89-120.

—— 1986. Work in Progress: Affect and Archetype. In *The Body in Analysis*. N. Schwartz-Salant and M. Stein, eds. Chiron Publications: Wilmette, IL. p. 183-203.

—— 1987a. "A brief report: affect and archetype." *Journal of Analytical Psychology* 32 (1): 35-46.

—— 1987b. Affect and Archetype in Analysis. In *Archetypal Processes in Psychotherapy*. N. Schwartz-Salant and M. Stein, eds. p. 131-162.

—— 1987c. Kinship Libido: Shadow in Marriage and Family. In *The Archetype of Shadow in a Split World*. M. A. Mattoon, ed. p. 387-399.

Stewart, L. H. and Stewart, C. T. 1979. Play, Games and Affects: A Contribution Toward a Comprehensive Theory of Play. In *Play as a Context*. A. T. Cheska, ed. Proceedings of the Association for the Anthropological Study of Play (TAASP). Leisure Press: Westpoint, NY. p. 42-52.

Thompson, W. 1981. *The Time Falling Bodies Take to Light*. St. Martin's Press: NY.

Tompkins, S. 1962. *Affect Imagery Consciousness*. Volume I. Springer Publishing Company: NY.

—— 1963. *Affect Imagery Consciousness.* Volume II. Springer Publishing Company: NY.

—— 1982. Affect Theory. In *Emotion in the Human Face.* 2nd edition, P. Ekman, ed. Cambridge University Press: Cambridge. p. 353-395.

Tucker, M. and Hirsh-Pasek, K. 1993. Systems and Language: Implications for Acquisition. In *A Dynamic Systems Approach to Development: Applications.* L. Smith and E. Thelen, eds. The MIT Press: Cambridge, MA. p. 359-384.

Ulanov, A. 1971. *The Feminine in Jungian Psychology.* Northwestern University Press: Evanston, IL.

Uttal, W. 1990. "On some two-way barriers between models and mechanisms." *Perception and Psychophysics* 48: 188-203.

Von Franz, M-L. 1975. *C. G. Jung: His Myth in Our Time.* Hodder and Stoughton: London.

—— 1980. *Alchemy.* Inner City Books: Toronto.

Walker, B. 1983. *The Woman's Encyclopedia of Myths and Secrets.* Harper & Row: San Francisco.

Weiner, H. 1969. *9 1/2 Mystics: The Kabbalah Today.* Collier Books (Macmillan): NY.

Wilhelm, R. 1967. *The I Ching or Book of Changes.* Cary Baynes, trans. Princeton University Press: Princeton, NY.

Winnicott, D. 1951/1975. Transitional Objects and Transitional Phenomena. In *Through Paediatrics to Psycho-analysis.* Basic Books: New York. p. 229-242.

—— 1966. "The location of cultural experience." *International Journal of Psycho-Analysis* 48: 368-372.

Yogi, Maharishi Mahesh. 1967. *On the Bhagavad-Gita.* Penguin Books: Harmondsworth.

Zak, M. 1991. "Terminal chaos for information processing in neurodynamics." *Biological Cybernetics* 64: 343-351.

INDEX

You might also enjoy reading these Jungian publications:

Lifting the Veil
by Jane Kamerling & Fred Gustafson
ISBN 978-1-926715-75-9

Becoming by Deldon Anne McNeely
ISBN 978-1-926715-12-4

The Creative Soul by Lawrence Staples
ISBN 978-0-9810344-4-7

Guilt with a Twist by Lawrence Staples
ISBN 978-0-9776076-4-8

Enemy, Cripple, Beggar by Erel Shalit
ISBN 978-0-9776076-7-9

The Cycle of Life by Erel Shalit
ISBN 978-1-926715-50-6

Divine Madness by John R. Haule
ISBN 978-1-926715-04-9

Farming Soul by Patricia Damery
ISBN 978-1-926715-01-8

The Motherline by Naomi Ruth Lowinsky
ISBN 978-0-9810344-6-1

The Sister From Below by Naomi Ruth Lowinsky
ISBN 978-0-9810344-2-3

Resurrecting the Unicorn by Bud Harris
ISBN 978-0-9810344-0-9

Phone Orders Welcomed
Credit Cards Accepted
In Canada & the U.S. call 1-800-228-9316
International call +1-831-238-7799
www.fisherkingpress.com

CPSIA information can be obtained at www.ICGtesting.com
Printed in the USA
LVOW082357280313

326481LV00002B/527/P